Mastering Fractions with Factors & Multiples

A Step-by-Step Instructional Workbook with Examples, Practice Worksheets and Answers

by

Stacy Otillio & Frank Otillio

CLAY MAZE.com

TABLE OF CONTENTS

This book was created to help students develop a solid understanding of fractions, but because factors and multiples are so essential to working with fractions, these concepts are covered first.

Part 1 - Factors & Multiples

Part 2 - Fractions

Factors and Divisibility Rules

Finding factors is an essential skill for working with fractions, and divisibility rules can help us find factors quickly and easily.

Factors

Factors of a number are numbers that divide evenly into the given number leaving no remainder.

For example, factors of 6 are numbers that divide evenly into 6.
1, 2, 3 and 6 all divide evenly into 6, so these are factors of 6.

Divisibility Rules

Divisibility rules can be used to help find factors of numbers.

A number is divisible by:

2 if the number is even (the last digit of the number ends in 0, 2, 4, 6 or 8)

3 if the sum of the digits in the number is divisible by 3

4 if the number formed by the last two digits is divisible by 4
If a number is not divisible by 2, then it is not divisible by 4.

5 if the last digit of the number is either 0 or 5

6 if the number is divisible by both 2 and 3

8 if the number formed by the last three digits is divisible by 8
If a number is not divisible by 4, then it is not divisible by 8.

9 if the sum of the digits in the number is divisible by 9
If a number is not divisible by 3, then it is not divisible by 9.

10 if the last digit of the number is 0

There is a rule for 7, but due to its complexity it's usually faster to just test by division.

Example: Use the divisibility rules to see which numbers (2 through 10) divide evenly into 1,280.

2 YES

1,280 is an even number and even numbers are divisible by 2.

3 NO

The sum of the digits in 1,280 is 1+2+8+0 = 11. 11 is NOT divisible by 3, so 1,280 is not divisible by 3.

4 YES

The last 2 digits of 1,280 form the number 80. 80 is divisible by 4, so 1,280 is divisible by 4.

5 YES

The last digit of 1,280 is 0, so it is divisible by 5.

6 NO

1,280 is not divisible by both 2 and 3, so it is not divisible by 6.

7 NO

Using division, 1,280 ÷ 7 = 182 with a remainder of 6, so it is not evenly divisible by 7 (since there is a remainder).

8 YES

The last 3 digits of 1,280 form the number 280. 280 is divisible by 8, so 1,280 is divisible by 8.

9 NO

The sum of the digits in 1,280 is 1+2+8+0 = 11. 11 is NOT divisible by 9, so 1,280 is not divisible by 9.

This test could have been skipped, since we already know that 1,280 is not divisible by 3. If a number is not divisible by 3, then it is not divisible by 9.

10 YES

The last digit of 1,280 is 0, so it is divisible by 10.

Name ______________________________ Date ______________

Divisibility Rules

Circle the numbers (2 through 10) that divide evenly into the given numbers.

1. 88

2 3 4 5 6 7 8 9 10

2. 56

2 3 4 5 6 7 8 9 10

3. 360

2 3 4 5 6 7 8 9 10

4. 786

2 3 4 5 6 7 8 9 10

5. 230

2 3 4 5 6 7 8 9 10

6. 225

2 3 4 5 6 7 8 9 10

7. 2,394

2 3 4 5 6 7 8 9 10

8. 4,130

2 3 4 5 6 7 8 9 10

9. 7,072

2 3 4 5 6 7 8 9 10

10. 12,168

2 3 4 5 6 7 8 9 10

Name ____________________ Date ____________________

Divisibility Rules

Circle the numbers (2 through 10) that divide evenly into the given numbers.

1. 45 2 3 4 5 6 7 8 9 10	2. 54 2 3 4 5 6 7 8 9 10
3. 174 2 3 4 5 6 7 8 9 10	4. 816 2 3 4 5 6 7 8 9 10
5. 728 2 3 4 5 6 7 8 9 10	6. 645 2 3 4 5 6 7 8 9 10
7. 5,635 2 3 4 5 6 7 8 9 10	8. 3,610 2 3 4 5 6 7 8 9 10
9. 9,846 2 3 4 5 6 7 8 9 10	10. 11,944 2 3 4 5 6 7 8 9 10

Squares and Square Roots

Some numbers have many factors and the process of finding them all can require a lot of divisibility testing. Using square roots and perfect squares can considerably reduce the amount of testing needed to find these factors by setting a limit on how high to test.

Squares

The **square** of a number is the result of multiplying that number by itself. It is written as the number with an **exponent** of 2.

For example, 3 squared is written as: 3^2

$$3^2 = 3 \times 3 = 9$$

Perfect Squares

A perfect square is a number that is the result of multiplying an integer by itself.

Integers are like the counting numbers, but they also include negative numbers and 0: { ..., -3, -2, -1, 0, 1, 2, 3, ... }
For our purposes, we're only concerned with positive numbers.

Other numbers like decimals and fractions can be squared, but **perfect squares are only squares of integers.**

Square Roots

Square roots are the opposite of squares. The **square root** of a number is the number which when squared equals the original number, and the radical symbol $\sqrt{}$ is used to represent it.

For example, the square root of 9 is 3, because $3^2 = 9$. More examples are shown below:

$\sqrt{9} = 3$ *(since $3^2 = 9$)* $\sqrt{16} = 4$ *(since $4^2 = 16$)*

$\sqrt{64} = 8$ *(since $8^2 = 64$)* $\sqrt{100} = 10$ *(since $10^2 = 100$)*

Consecutive Integers and Consecutive Perfect Squares

Consecutive integers are integers that follow one after another, like 3 and 4.

There are no integers between any two consecutive integers.
For example, there are no integers between 3 and 4. This doesn't mean there are no numbers between 3 and 4, just that there are no integers between them.

Consecutive perfect squares are squares of consecutive integers. These are numbers like 3^2 and 4^2 or 6^2 and 7^2.

There are no perfect squares between any two consecutive perfect squares.
For example, there are no perfect squares between 2^2 and 3^2 or between 5^2 and 6^2. There are integers and other numbers between them, but not perfect squares.

Numbers like 9 and 16 are consecutive perfect squares, because $3^2 = 9$ and $4^2 = 16$, and 3 and 4 are consecutive integers.

Below is a chart of the integers 1 - 12 in the first row and their squares in the second row.

1	2	3	4	5	6	7	8	9	10	11	12
1	4	9	16	25	36	49	64	81	100	121	144

Example: Find two consecutive perfect squares that the given numbers are between and write the square root of the lower one in the grey box. Use the chart above for reference.

a. 10

From the chart, we can see that 10 lies between the consecutive squares: 9 and 16.
($9 = 3^2$ and $16 = 4^2$)

$\sqrt{9} = 3$ $\quad$ $\underline{9} < 10 < \underline{16}$

b. 95

From the chart, we can see that 95 lies between the consecutive squares: 81 and 100.
($81 = 9^2$ and $100 = 10^2$)

$\sqrt{81} = 9$ $\quad$ $\underline{81} < 95 < \underline{100}$

This concept will become useful in the next section when determining how high to test for factors by limiting the amount of numbers needed to test.

Name ______________________ Date ______________

Squares and Square Roots

Fill in the chart with the squares for numbers 1 through 12 to use with the problems below.

1	2	3	4	5	6	7	8	9	10	11	12

Fill in the blanks with the two consecutive perfect squares that the given numbers are between and write the **square roots** of the lower ones in the grey boxes. Use the chart for reference.

	Square root					
1.	$\sqrt{16} = 4$ 4	16	<	20	<	25
2.		____	<	40	<	____
3.		____	<	48	<	____
4.		____	<	52	<	____
5.		____	<	65	<	____
6.		____	<	80	<	____
7.		____	<	95	<	____
8.		____	<	132	<	____

Finding Factors

There are three main steps to follow when finding the factors of a number.

Step 1: Find out how high to test numbers for divisibility.
- If the given number is a perfect square, we only need to test up to its square root.
- If the given number is not a perfect square, we only need to test up to the square root of the closest perfect square that's **less than** the given number.

Step 2: Use the divisibility rules to test for factors up to the number found in step 1. These will make up the first half of the factors.

Step 3: Divide the given number by the factors found in step 2 to find the rest of the factors.

Some numbers have only two factors: itself and 1. These are called **prime numbers**.

Example: Find the factors of 63.

Step 1: Figure out how high to test for divisibility.

63 is not a perfect square. The closest perfect square that's less than 63 is 49.
The square root of 49 is 7, so we only need to test up to 7.

Step 2: Use the divisibility rules (up to 7) to test the divisibility of 63.

✓ **1**: **1 is a factor**, since 1 will always divide evenly into an integer.

2: 63 is not an even number, so it is NOT divisible by 2.

✓ **3**: The sum of the digits in 63 is 6 + 3 = 9, which is divisible by 3, so **3 is a factor**.

4: 63 is not divisible by 2, so it is NOT divisible by 4.

5: 63 does not end in a 0 or a 5, so it is NOT divisible by 5.

6: 63 is not divisible by both 2 and 3, so it is NOT divisible by 6.

✓ **7**: 63 is divisible by 7 (7 x 9 = 63), so **7 is a factor**.

The first half of the factors of 63 are: 1, 3, and 7.

Step 3: Divide 63 by the factors found in **Step 2** (1, 3, and 7) to find the rest of the factors.

1: 63÷1=63 Factors: 1, 63

3: 63÷3=21 Factors: 3, 21

7: 63÷7=9 Factors: 7, 9

The factors of 63 are: 1, 3, 7, 9, 21 and 63

Name ______________________ Date ______________

Find the Factors

List the factors of the numbers below. If a given number is prime, write prime.

1. 8	2. 15
3. 17	4. 28
5. 32	6. 40
7. 65	8. 100
9. 124	10. 130

Name ______________________________ Date ______________

Find the Factors

List the factors of the numbers below. If a given number is prime, write prime.

1. 9	2. 24
3. 36	4. 45
5. 51	6. 88
7. 85	8. 113
9. 115	10. 135

Name ______________________ Date ______________

Find the Factors

List the factors of the numbers below. If a given number is prime, write prime.

1. 10	2. 25
3. 42	4. 70
5. 84	6. 91
7. 110	8. 116
9. 127	10. 138

Greatest Common Factor (GCF)

The **Greatest Common Factor** (GCF) of two or more numbers is the largest factor that those numbers have in common.

To find the GCF, first find the factors of each number. The largest factor that is shared by the given numbers is the GCF.

Example: Find the GCF of 20 and 35.

Find the factors of each number.

Factors of 20:	1	2	4	5	10	20
Factors of 35:	1	5	7	35		

5 is the largest factor that both numbers have in common, so **the GCF of 20 and 35 is 5.**

This same method can be used for more than 2 numbers.

Example: Find the GCF of 8, 16 and 48.

Find the factors of each number.

Factors of 8:	1	2	4	8						
Factors of 16:	1	2	4	8	16					
Factors of 48:	1	2	3	4	6	8	12	16	24	48

8 is the largest factor that all three numbers have in common, so their **GCF is 8.**

As a shortcut, see if the smallest of the given numbers divides evenly into the others. If so, that number is the GCF.

Name ______________________ Date ______________

GCF (Greatest Common Factor)

Find the GCF.

1. 6, 8

GCF: ________

2. 10, 55

GCF: ________

3. 13, 26

GCF: ________

4. 15, 65

GCF: ________

5. 16, 22

GCF: ________

6. 24, 39

GCF: ________

7. 30, 45

GCF: ________

8. 49, 56

GCF: ________

9. 8, 36, 56

GCF: ________

10. 22, 55, 88

GCF: ________

Name ______________________ Date ______________

GCF (Greatest Common Factor)

Find the GCF.

1. 6, 24

GCF: ________

2. 14, 28

GCF: ________

3. 15, 45

GCF: ________

4. 21, 28

GCF: ________

5. 24, 36

GCF: ________

6. 30, 42

GCF: ________

7. 50, 125

GCF: ________

8. 56, 120

GCF: ________

9. 3, 5, 11

GCF: ________

10. 12, 36, 72

GCF: ________

Name ______________________ Date ______________

GCF (Greatest Common Factor)

Find the GCF.

1. 8, 12

GCF: ________

2. 15, 33

GCF: ________

3. 21, 35

GCF: ________

4. 24, 90

GCF: ________

5. 34, 39

GCF: ________

6. 35, 40

GCF: ________

7. 48, 96

GCF: ________

8. 70, 140

GCF: ________

9. 12, 24, 45

GCF: ________

10. 30, 48, 54

GCF: ________

Name ______________________________ Date ____________________

GCF (Greatest Common Factor)

Find the GCF.

1. 12, 15

GCF: ________

2. 13, 39

GCF: ________

3. 16, 18

GCF: ________

4. 21, 27

GCF: ________

5. 22, 121

GCF: ________

6. 26, 34

GCF: ________

7. 45, 55

GCF: ________

8. 36, 54

GCF: ________

9. 15, 25, 55

GCF: ________

10. 24, 36, 60

GCF: ________

Least Common Multiple (LCM)

Multiples

A **multiple** of a number is the result of multiplying that number by an integer.

Below are some examples of multiples of 3 and multiples of 5 (15 is a common multiple).

Multiples of 3	**Multiples of 5**
3 x 1 = 3	5 x 1 = 5
3 x 2 = 6	5 x 2 = 10
3 x 3 = 9	5 x 3 = 15
3 x 4 = 12	5 x 4 = 20
3 x 5 = 15	5 x 5 = 25

If a smaller number divides evenly into a larger number, then that larger number is a multiple of the smaller number.

Common Multiples

To find **common multiples** of two numbers, first find multiples of the larger of the given numbers, then test to see which ones are also multiples of the smaller number. They are multiples of the smaller number if they are divisible by the smaller number.

Example: Find the first three common multiples of 2 and 3.

First, find multiples of the larger number (3), then see which of those multiples are also multiples of 2 until three are found. A number is a multiple of 2 if it is divisible by 2.

Multiples of 3	3	6	9	12	15	18
Multiples of 2		✓		✓		✓

3, 9 and 15 are not multiples of 2, since they are not divisible by 2.

The first three common multiples of 2 and 3 are: 6, 12 and 18.

The smallest multiple of those numbers is 6. This smallest common multiple is called the Least Common Multiple (or ***LCM****). It can also be referred to as the Lowest Common Multiple.*

Least Common Multiple

The **Least Common Multiple (LCM)** of two numbers is the smallest number that both of the given numbers divide evenly into. This is the smallest common multiple of the numbers.

To find the least common multiple of two numbers:

A **If the larger of the two numbers is evenly divisible by the smaller one**, then that larger number is the **LCM**.

B **If the larger of the two numbers is NOT evenly divisible by the smaller one**, then find multiples of the larger number until you find the first one that is also a multiple of the smaller number. This first common multiple is the smallest common multiple, which is the **LCM**.

Example: Find the LCM of 4 and 12.

A: See if the smaller number divides evenly into the larger number.

12 is divisible by 4, so **12 is the LCM.**

LCM(4, 12) = 12

Example: Find the LCM of 8 and 10.

A: See if the smaller number divides evenly into the larger number.
10 is NOT divisible by 8, so 10 is NOT the LCM.

B: Find multiples of the larger number (10), until we find one that is also a multiple of the smaller number (8).

10x2 = 20	20 is not divisible by 8, so it's NOT a multiple of 8.
10x3 = 30	30 is not divisible by 8, so it's NOT a multiple of 8.
10x4 = 40	40 is divisible by 8, so it's a multiple of 8 (8 x 5 = 40). It is the smallest multiple of 10 that is also divisible by 8, so 40 is the LCM of 8 and 10.

LCM(8, 10) = 40

Name ______________________ Date ______________

LCM (Least Common Multiple)

Find the LCM.

1. 2, 4

LCM: ________

2. 4, 6

LCM: ________

3. 5, 11

LCM: ________

4. 6, 8

LCM: ________

5. 6, 10

LCM: ________

6. 5, 12

LCM: ________

7. 9, 15

LCM: ________

8. 9, 12

LCM: ________

9. 12, 30

LCM: ________

10. 16, 24

LCM: ________

Name ______________________ Date ______________

LCM (Least Common Multiple)

Find the LCM.

1. 2, 3

LCM: ________

2. 3, 7

LCM: ________

3. 4, 10

LCM: ________

4. 6, 12

LCM: ________

5. 6, 20

LCM: ________

6. 8, 9

LCM: ________

7. 8, 11

LCM: ________

8. 10, 25

LCM: ________

9. 12, 15

LCM: ________

10. 15, 45

LCM: ________

Name ______________________ Date ______________

LCM (Least Common Multiple)

Find the LCM.

1. 2, 5

LCM: ________

2. 3, 25

LCM: ________

3. 4, 50

LCM: ________

4. 5, 9

LCM: ________

5. 9, 21

LCM: ________

6. 6, 15

LCM: ________

7. 10, 15

LCM: ________

8. 10, 45

LCM: ________

9. 12, 20

LCM: ________

10. 15, 25

LCM: ________

Name ______________________________ Date ______________

LCM (Least Common Multiple)

Find the LCM.

1. 2, 7

   LCM: ________

2. 4, 8

   LCM: ________

3. 5, 12

   LCM: ________

4. 6, 9

   LCM: ________

5. 8, 12

   LCM: ________

6. 10, 12

   LCM: ________

7. 12, 16

   LCM: ________

8. 8, 20

   LCM: ________

9. 15, 30

   LCM: ________

10. 20, 25

    LCM: ________

Fractions and Equivalent Fractions

Fractions Review

A **fraction** represents an amount out of a whole. If we think of a whole amount of something as the sum of equal parts, a fraction can be used to represent 1 or more of those equal parts.

For example, suppose there is a whole pizza cut into 6 equal slices.

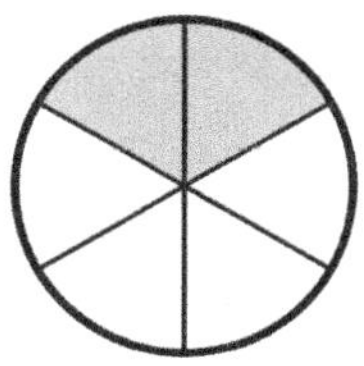

If you ate 2 of those 6 slices, then you ate $\frac{2}{6}$ of the pizza.

The top number (2) is called the **numerator** and represents the amount eaten.

The bottom number (6) is called the **denominator** and is the total number of equal slices that the pizza is divided into.

Equivalent Fractions

Equivalent fractions are fractions that have the same value with different numerators and denominators.

For example, suppose you have 2 pizzas. One is cut into 6 equal slices and the other is cut into 3 equal slices.

Eating 2 slices from the first pizza (cut into 6 equal pieces) is equivalent to eating 1 slice from the second pizza (cut into 3 equal pieces). The total amount shaded in for each pizza is the same size. It's the same amount, but a different number of slices.

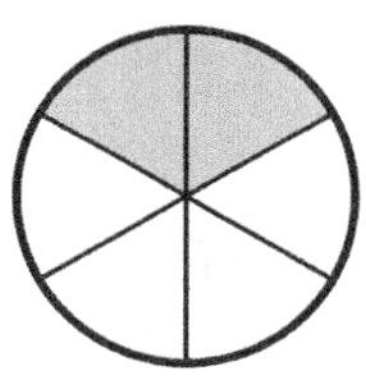

$\frac{2}{6}$ is equivalent to $\frac{1}{3}$

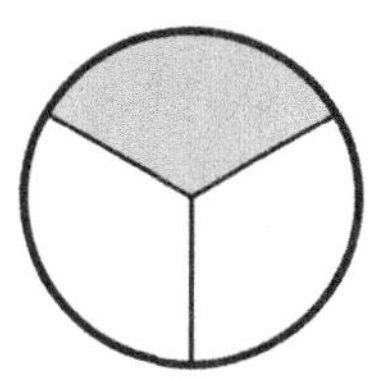

The fractions $\frac{2}{6}$ and $\frac{1}{3}$ are **equivalent fractions**.

They represent the same amount, but they have different numerators and denominators.

Finding Equivalent Fractions

A fraction will have the same value if you multiply (or divide) BOTH the numerator and the denominator by the same number. This is how we can find equivalent fractions.

Multiplying: $\frac{1\ (\times 2)}{2\ (\times 2)} = \frac{2}{4}$ $\frac{1}{2}$ and $\frac{2}{4}$ are equivalent fractions.

Dividing: $\frac{5\ (\div 5)}{15\ (\div 5)} = \frac{1}{3}$ $\frac{5}{15}$ and $\frac{1}{3}$ are equivalent fractions.

Make sure that BOTH the numerator and the denominator are multiplied by the SAME number or divided by the SAME number.

Example: Find the missing values that make the fractions equivalent.

a. $\frac{3}{5} = \frac{?}{15}$

The denominators are known in both fractions: 5 and 15

Multiplying by 3 would change the denominator of 5 into 15, so if both the numerator and the denominator of the first fraction are multiplied by 3, we can get an equivalent fraction with 15 as the denominator.

$\frac{3\ (\times 3)}{5\ (\times 3)} = \frac{9}{15}$ The missing numerator is 9.

b. $\frac{3}{?} = \frac{21}{28}$

The numerators are known in both fractions: 3 and 21

Dividing by 7 would change the numerator of 21 into 3, so if both the numerator and the denominator of the second fraction are divided by 7, we can get an equivalent fraction with 3 as the numerator.

$\frac{3}{4} = \frac{21\ (\div 7)}{28\ (\div 7)}$ The missing denominator is 4.

Name ______________________________ Date ______________________

Equivalent Fractions

Fill in each blank with the missing numerator or denominator to make the fractions equivalent.

1. $\frac{2}{\underline{\quad}} = \frac{18}{81}$
2. $\frac{5}{6} = \frac{\underline{\quad}}{24}$
3. $\frac{7}{\underline{\quad}} = \frac{49}{70}$
4. $\frac{1}{\underline{\quad}} = \frac{3}{6}$
5. $\frac{3}{9} = \frac{\underline{\quad}}{63}$
6. $\frac{\underline{\quad}}{8} = \frac{2}{16}$
7. $\frac{9}{25} = \frac{\underline{\quad}}{100}$
8. $\frac{2}{\underline{\quad}} = \frac{4}{12}$
9. $\frac{1}{\underline{\quad}} = \frac{4}{20}$
10. $\frac{2}{8} = \frac{10}{\underline{\quad}}$
11. $\frac{\underline{\quad}}{10} = \frac{8}{20}$
12. $\frac{5}{8} = \frac{25}{\underline{\quad}}$
13. $\frac{\underline{\quad}}{7} = \frac{35}{49}$
14. $\frac{\underline{\quad}}{12} = \frac{12}{24}$
15. $\frac{3}{4} = \frac{6}{\underline{\quad}}$
16. $\frac{7}{12} = \frac{\underline{\quad}}{36}$
17. $\frac{1}{9} = \frac{9}{\underline{\quad}}$
18. $\frac{\underline{\quad}}{10} = \frac{45}{50}$
19. $\frac{\underline{\quad}}{11} = \frac{28}{44}$
20. $\frac{9}{12} = \frac{18}{\underline{\quad}}$
21. $\frac{8}{\underline{\quad}} = \frac{48}{66}$
22. $\frac{6}{8} = \frac{60}{\underline{\quad}}$
23. $\frac{\underline{\quad}}{100} = \frac{10}{200}$
24. $\frac{4}{\underline{\quad}} = \frac{40}{50}$
25. $\frac{3}{\underline{\quad}} = \frac{21}{84}$
26. $\frac{4}{12} = \frac{\underline{\quad}}{36}$
27. $\frac{8}{9} = \frac{\underline{\quad}}{63}$
28. $\frac{\underline{\quad}}{11} = \frac{24}{88}$
29. $\frac{11}{12} = \frac{55}{\underline{\quad}}$
30. $\frac{8}{12} = \frac{32}{\underline{\quad}}$

Name ______________________________ Date ______________

Equivalent Fractions

Fill in each blank with the missing numerator or denominator to make the fractions equivalent.

1. $\frac{3}{7} = \frac{12}{\underline{\quad}}$
2. $\frac{2}{3} = \frac{14}{\underline{\quad}}$
3. $\frac{4}{\underline{\quad}} = \frac{28}{42}$
4. $\frac{1}{3} = \frac{\underline{\quad}}{15}$
5. $\frac{7}{8} = \frac{42}{\underline{\quad}}$
6. $\frac{\underline{\quad}}{4} = \frac{20}{40}$
7. $\frac{10}{\underline{\quad}} = \frac{50}{55}$
8. $\frac{\underline{\quad}}{2} = \frac{4}{8}$
9. $\frac{6}{50} = \frac{12}{\underline{\quad}}$
10. $\frac{2}{8} = \frac{\underline{\quad}}{16}$
11. $\frac{\underline{\quad}}{11} = \frac{64}{88}$
12. $\frac{1}{5} = \frac{\underline{\quad}}{40}$
13. $\frac{\underline{\quad}}{25} = \frac{16}{50}$
14. $\frac{\underline{\quad}}{9} = \frac{28}{36}$
15. $\frac{2}{\underline{\quad}} = \frac{6}{36}$
16. $\frac{9}{12} = \frac{36}{\underline{\quad}}$
17. $\frac{3}{\underline{\quad}} = \frac{15}{55}$
18. $\frac{3}{\underline{\quad}} = \frac{6}{20}$
19. $\frac{4}{100} = \frac{\underline{\quad}}{400}$
20. $\frac{6}{12} = \frac{\underline{\quad}}{60}$
21. $\frac{3}{8} = \frac{\underline{\quad}}{56}$
22. $\frac{7}{12} = \frac{\underline{\quad}}{24}$
23. $\frac{4}{7} = \frac{8}{\underline{\quad}}$
24. $\frac{5}{8} = \frac{15}{\underline{\quad}}$
25. $\frac{\underline{\quad}}{11} = \frac{4}{22}$
26. $\frac{10}{12} = \frac{100}{\underline{\quad}}$
27. $\frac{4}{\underline{\quad}} = \frac{32}{72}$
28. $\frac{\underline{\quad}}{9} = \frac{30}{45}$
29. $\frac{8}{\underline{\quad}} = \frac{72}{81}$
30. $\frac{5}{10} = \frac{\underline{\quad}}{100}$

Name ____________________ Date ____________________

Equivalent Fractions

Fill in each blank with the missing numerator or denominator to make the fractions equivalent.

1. $\frac{6}{10} = \frac{\_\_}{40}$

2. $\frac{\_\_}{6} = \frac{35}{42}$

3. $\frac{4}{11} = \frac{28}{\_\_}$

4. $\frac{5}{\_\_} = \frac{15}{75}$

5. $\frac{1}{4} = \frac{\_\_}{8}$

6. $\frac{2}{\_\_} = \frac{8}{36}$

7. $\frac{\_\_}{2} = \frac{4}{8}$

8. $\frac{4}{9} = \frac{20}{\_\_}$

9. $\frac{\_\_}{7} = \frac{6}{21}$

10. $\frac{5}{8} = \frac{30}{\_\_}$

11. $\frac{3}{\_\_} = \frac{6}{10}$

12. $\frac{10}{12} = \frac{\_\_}{60}$

13. $\frac{2}{11} = \frac{10}{\_\_}$

14. $\frac{\_\_}{10} = \frac{48}{60}$

15. $\frac{3}{\_\_} = \frac{30}{40}$

16. $\frac{\_\_}{12} = \frac{20}{60}$

17. $\frac{8}{9} = \frac{\_\_}{36}$

18. $\frac{5}{12} = \frac{60}{\_\_}$

19. $\frac{1}{\_\_} = \frac{8}{24}$

20. $\frac{11}{12} = \frac{33}{\_\_}$

21. $\frac{\_\_}{7} = \frac{10}{14}$

22. $\frac{8}{25} = \frac{\_\_}{50}$

23. $\frac{\_\_}{5} = \frac{12}{30}$

24. $\frac{7}{\_\_} = \frac{77}{88}$

25. $\frac{\_\_}{6} = \frac{15}{30}$

26. $\frac{2}{\_\_} = \frac{6}{12}$

27. $\frac{3}{8} = \frac{\_\_}{32}$

28. $\frac{6}{7} = \frac{48}{\_\_}$

29. $\frac{2}{3} = \frac{\_\_}{6}$

30. $\frac{7}{9} = \frac{35}{\_\_}$

Name ______________________________ Date ____________________

Equivalent Fractions

Fill in each blank with the missing numerator or denominator to make the fractions equivalent.

1. $\frac{4}{8} = \frac{\_\_\_}{64}$
2. $\frac{1}{10} = \frac{10}{\_\_\_}$
3. $\frac{3}{7} = \frac{\_\_\_}{56}$
4. $\frac{1}{\_\_\_} = \frac{8}{56}$
5. $\frac{2}{\_\_\_} = \frac{6}{9}$
6. $\frac{\_\_\_}{4} = \frac{4}{16}$
7. $\frac{2}{\_\_\_} = \frac{14}{35}$
8. $\frac{4}{7} = \frac{20}{\_\_\_}$
9. $\frac{5}{\_\_\_} = \frac{40}{72}$
10. $\frac{7}{10} = \frac{\_\_\_}{60}$
11. $\frac{3}{\_\_\_} = \frac{33}{55}$
12. $\frac{7}{11} = \frac{49}{\_\_\_}$
13. $\frac{4}{\_\_\_} = \frac{24}{60}$
14. $\frac{3}{8} = \frac{\_\_\_}{80}$
15. $\frac{4}{8} = \frac{\_\_\_}{32}$
16. $\frac{\_\_\_}{11} = \frac{25}{55}$
17. $\frac{5}{6} = \frac{15}{\_\_\_}$
18. $\frac{\_\_\_}{5} = \frac{27}{45}$
19. $\frac{6}{8} = \frac{\_\_\_}{16}$
20. $\frac{5}{\_\_\_} = \frac{20}{32}$
21. $\frac{2}{6} = \frac{10}{\_\_\_}$
22. $\frac{4}{25} = \frac{16}{\_\_\_}$
23. $\frac{\_\_\_}{9} = \frac{21}{63}$
24. $\frac{\_\_\_}{12} = \frac{77}{84}$
25. $\frac{\_\_\_}{9} = \frac{16}{18}$
26. $\frac{9}{12} = \frac{\_\_\_}{36}$
27. $\frac{\_\_\_}{12} = \frac{24}{48}$
28. $\frac{7}{\_\_\_} = \frac{28}{48}$
29. $\frac{8}{11} = \frac{\_\_\_}{33}$
30. $\frac{9}{10} = \frac{\_\_\_}{80}$

Comparing Fractions

Comparing Fractions with the Same Denominators

When two fractions have the same denominator, we can tell which one is bigger by looking at the numerator, since the numerators tell how many parts out of the whole. The fraction with the higher numerator is the larger fraction, because it indicates more parts.

Consider the pizza example, where the pizza is cut into 6 equal slices. We can tell right away that 4 of those 6 slices would be more than 2 of the 6 slices, since 4 is more than 2.

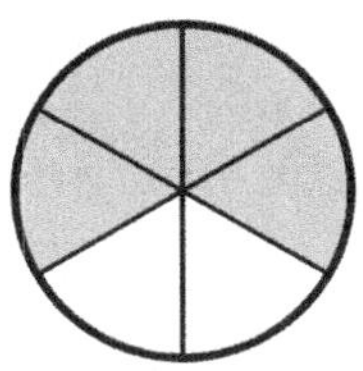

$\frac{4}{6}$

>

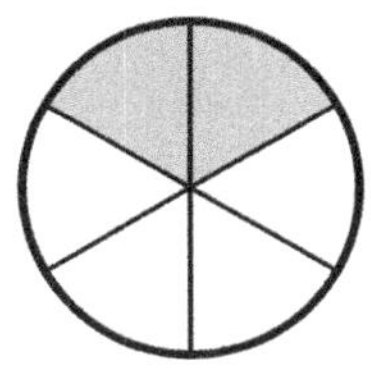

$\frac{2}{6}$

$$\frac{4}{6} > \frac{2}{6}$$

Example: Compare the fractions.

$\frac{2}{10}$ and $\frac{8}{10}$

These fractions have the same denominator, so only the numerators need to be compared. 2 is less than 8, so $\frac{2}{10}$ is less than $\frac{8}{10}$.

$$\frac{2}{10} < \frac{8}{10}$$

Comparing Fractions with Different Denominators

It's often not obvious which fraction is larger or smaller with different denominators.

For example: $\frac{4}{6}$ and $\frac{7}{11}$. Can you tell which one is larger?

To compare fractions with different denominators, we can turn them into equivalent fractions with the same denominators then compare them.

There is a shortcut based off of this concept, which we cover next.

To compare fractions with different denominators, place them side by side and follow these steps:

Step 1: Multiply the denominator of the second fraction by the numerator of the first and write that number near the first fraction.

Step 2: Multiply the denominator of the first fraction by the numerator of the second and write that number near the second fraction.

Step 3: Compare the two numbers.

- If the number by the first fraction is larger, then the first fraction is larger.
- If the number by the second fraction is larger, then the second fraction is larger.

Example: Which fraction is larger: $\frac{4}{6}$ or $\frac{7}{11}$?

Step 1: Multiply the denominator of the second fraction by the numerator of the first and write that number near the first fraction.

44 $\frac{4}{6}$ $\frac{7}{11}$

$11 \times 4 = 44$

Step 2: Multiply the denominator of the first fraction by the numerator of the second and write that number near the second fraction.

44 $\frac{4}{6}$ $\frac{7}{11}$ 42

$6 \times 7 = 42$ $11 \times 4 = 44$

Step 3: Compare the two numbers next to each fraction.

44 > 42, so the first fraction is greater than the second fraction.

$$\frac{4}{6} > \frac{7}{11}$$

Why it Works

The reason this method works is because we're really just changing the fractions into two equivalent ones with the same denominator. This new denominator is the product of the two original denominators, so each fraction's numerator and denominator are multiplied by the other fraction's denominator.

$$\frac{4\ (\times 11)}{6\ (\times 11)} = \frac{44}{66} \quad \text{and} \quad \frac{7\ (\times 6)}{11\ (\times 6)} = \frac{42}{66} \qquad \frac{44}{66} > \frac{42}{66} \quad \text{so} \quad \frac{4}{6} > \frac{7}{11}$$

Since we know both denominators will be the same and we're really only interested in comparing numerators, we don't have to carry out the multiplication on the denominators.

Name ______________________ Date ______________

Comparing Fractions

Compare the fractions and fill in the blanks with < or >.

1. $\frac{5}{6} \;\square\; \frac{2}{11}$	2. $\frac{2}{3} \;\square\; \frac{4}{8}$	3. $\frac{3}{6} \;\square\; \frac{6}{10}$
4. $\frac{8}{11} \;\square\; \frac{6}{7}$	5. $\frac{7}{9} \;\square\; \frac{2}{3}$	6. $\frac{6}{8} \;\square\; \frac{3}{7}$
7. $\frac{7}{9} \;\square\; \frac{5}{8}$	8. $\frac{8}{9} \;\square\; \frac{5}{7}$	9. $\frac{3}{9} \;\square\; \frac{2}{8}$
10. $\frac{4}{6} \;\square\; \frac{1}{4}$	11. $\frac{6}{12} \;\square\; \frac{5}{8}$	12. $\frac{2}{6} \;\square\; \frac{5}{7}$
13. $\frac{3}{5} \;\square\; \frac{6}{11}$	14. $\frac{5}{9} \;\square\; \frac{5}{6}$	15. $\frac{5}{8} \;\square\; \frac{3}{12}$
16. $\frac{5}{12} \;\square\; \frac{2}{4}$	17. $\frac{6}{10} \;\square\; \frac{6}{9}$	18. $\frac{4}{5} \;\square\; \frac{8}{9}$
19. $\frac{1}{4} \;\square\; \frac{4}{5}$	20. $\frac{7}{11} \;\square\; \frac{2}{6}$	21. $\frac{4}{10} \;\square\; \frac{5}{8}$
22. $\frac{3}{8} \;\square\; \frac{7}{10}$	23. $\frac{2}{3} \;\square\; \frac{9}{12}$	24. $\frac{4}{7} \;\square\; \frac{3}{9}$
25. $\frac{3}{10} \;\square\; \frac{1}{11}$	26. $\frac{9}{12} \;\square\; \frac{8}{10}$	27. $\frac{7}{9} \;\square\; \frac{5}{6}$
28. $\frac{1}{4} \;\square\; \frac{4}{7}$	29. $\frac{4}{5} \;\square\; \frac{3}{8}$	30. $\frac{4}{7} \;\square\; \frac{4}{11}$

Name ______________________ Date ______________

Comparing Fractions

Compare the fractions and fill in the blanks with < or >.

1. $\frac{1}{3} \square \frac{9}{12}$
2. $\frac{7}{10} \square \frac{2}{9}$
3. $\frac{5}{12} \square \frac{5}{6}$
4. $\frac{5}{7} \square \frac{1}{10}$
5. $\frac{3}{8} \square \frac{2}{6}$
6. $\frac{3}{7} \square \frac{2}{5}$
7. $\frac{3}{10} \square \frac{3}{6}$
8. $\frac{7}{9} \square \frac{9}{10}$
9. $\frac{3}{4} \square \frac{7}{11}$
10. $\frac{2}{3} \square \frac{7}{12}$
11. $\frac{5}{8} \square \frac{3}{4}$
12. $\frac{2}{7} \square \frac{7}{12}$
13. $\frac{3}{9} \square \frac{5}{7}$
14. $\frac{8}{10} \square \frac{2}{12}$
15. $\frac{9}{12} \square \frac{7}{9}$
16. $\frac{1}{4} \square \frac{2}{9}$
17. $\frac{5}{9} \square \frac{7}{8}$
18. $\frac{7}{8} \square \frac{3}{5}$
19. $\frac{1}{8} \square \frac{4}{7}$
20. $\frac{7}{12} \square \frac{7}{11}$
21. $\frac{2}{5} \square \frac{7}{10}$
22. $\frac{3}{8} \square \frac{2}{7}$
23. $\frac{2}{5} \square \frac{3}{8}$
24. $\frac{2}{3} \square \frac{8}{9}$
25. $\frac{2}{12} \square \frac{5}{6}$
26. $\frac{3}{5} \square \frac{5}{7}$
27. $\frac{4}{8} \square \frac{5}{6}$
28. $\frac{6}{7} \square \frac{4}{5}$
29. $\frac{3}{4} \square \frac{5}{6}$
30. $\frac{2}{9} \square \frac{6}{10}$

Name ____________________ Date ____________________

Comparing Fractions

Compare the fractions and fill in the blanks with < or >.

1. $\frac{7}{10}$ ☐ $\frac{6}{11}$
2. $\frac{6}{8}$ ☐ $\frac{2}{3}$
3. $\frac{1}{5}$ ☐ $\frac{3}{7}$
4. $\frac{7}{9}$ ☐ $\frac{7}{12}$
5. $\frac{2}{7}$ ☐ $\frac{2}{4}$
6. $\frac{8}{10}$ ☐ $\frac{5}{12}$
7. $\frac{4}{5}$ ☐ $\frac{6}{7}$
8. $\frac{9}{12}$ ☐ $\frac{8}{11}$
9. $\frac{3}{4}$ ☐ $\frac{5}{8}$
10. $\frac{1}{6}$ ☐ $\frac{1}{5}$
11. $\frac{4}{5}$ ☐ $\frac{5}{8}$
12. $\frac{3}{7}$ ☐ $\frac{4}{9}$
13. $\frac{8}{11}$ ☐ $\frac{8}{9}$
14. $\frac{8}{10}$ ☐ $\frac{1}{2}$
15. $\frac{4}{7}$ ☐ $\frac{3}{12}$
16. $\frac{5}{9}$ ☐ $\frac{3}{12}$
17. $\frac{5}{6}$ ☐ $\frac{7}{8}$
18. $\frac{5}{12}$ ☐ $\frac{6}{8}$
19. $\frac{6}{8}$ ☐ $\frac{5}{7}$
20. $\frac{9}{11}$ ☐ $\frac{3}{5}$
21. $\frac{7}{9}$ ☐ $\frac{8}{10}$
22. $\frac{4}{6}$ ☐ $\frac{8}{9}$
23. $\frac{3}{8}$ ☐ $\frac{2}{6}$
24. $\frac{3}{4}$ ☐ $\frac{6}{11}$
25. $\frac{7}{12}$ ☐ $\frac{9}{11}$
26. $\frac{6}{7}$ ☐ $\frac{2}{8}$
27. $\frac{9}{10}$ ☐ $\frac{6}{7}$
28. $\frac{3}{6}$ ☐ $\frac{1}{5}$
29. $\frac{1}{4}$ ☐ $\frac{2}{5}$
30. $\frac{3}{6}$ ☐ $\frac{4}{5}$

Name ______________________________ Date ______________

Comparing Fractions

Compare the fractions and fill in the blanks with < or >.

1. $\frac{3}{5}$ $\square$ $\frac{2}{4}$	2. $\frac{6}{8}$ $\square$ $\frac{6}{10}$	3. $\frac{3}{12}$ $\square$ $\frac{4}{8}$
4. $\frac{9}{11}$ $\square$ $\frac{7}{12}$	5. $\frac{3}{7}$ $\square$ $\frac{2}{9}$	6. $\frac{7}{8}$ $\square$ $\frac{5}{12}$
7. $\frac{2}{6}$ $\square$ $\frac{8}{9}$	8. $\frac{9}{12}$ $\square$ $\frac{8}{9}$	9. $\frac{1}{5}$ $\square$ $\frac{7}{9}$
10. $\frac{5}{9}$ $\square$ $\frac{4}{7}$	11. $\frac{1}{4}$ $\square$ $\frac{1}{5}$	12. $\frac{2}{6}$ $\square$ $\frac{1}{4}$
13. $\frac{4}{5}$ $\square$ $\frac{7}{12}$	14. $\frac{1}{3}$ $\square$ $\frac{3}{8}$	15. $\frac{3}{7}$ $\square$ $\frac{3}{9}$
16. $\frac{3}{8}$ $\square$ $\frac{5}{7}$	17. $\frac{4}{5}$ $\square$ $\frac{9}{10}$	18. $\frac{3}{5}$ $\square$ $\frac{4}{7}$
19. $\frac{9}{10}$ $\square$ $\frac{3}{4}$	20. $\frac{7}{9}$ $\square$ $\frac{1}{6}$	21. $\frac{2}{3}$ $\square$ $\frac{8}{10}$
22. $\frac{2}{8}$ $\square$ $\frac{6}{7}$	23. $\frac{3}{11}$ $\square$ $\frac{6}{9}$	24. $\frac{2}{9}$ $\square$ $\frac{5}{11}$
25. $\frac{8}{12}$ $\square$ $\frac{3}{8}$	26. $\frac{6}{9}$ $\square$ $\frac{8}{11}$	27. $\frac{5}{6}$ $\square$ $\frac{8}{9}$
28. $\frac{8}{11}$ $\square$ $\frac{8}{10}$	29. $\frac{8}{10}$ $\square$ $\frac{8}{12}$	30. $\frac{4}{10}$ $\square$ $\frac{2}{8}$

Simplifying Fractions

Simplified Fractions

A **simplified fraction** has a numerator and a denominator that have no common factors other than 1.

To simplify a fraction:

Step 1: Find the GCF (Greatest Common Factor) of the numerator and the denominator.

Step 2: Find an equivalent fraction by dividing the numerator and denominator by their GCF found in Step 1.

Example: Simplify the fraction: $\frac{12}{16}$

Step 1: The GCF of 12 and 16 is 4.

Step 2: Divide the numerator and denominator by their GCF (4) to get the simplified fraction.

$$\frac{12\,(\div 4)}{16\,(\div 4)} = \frac{3}{4}$$

If the GCF isn't immediately apparent, the numerator and the denominator can be incrementally divided by common factors until there are no common factors left.

Example: Simplify the fraction: $\frac{165}{180}$

The GCF of 165 and 180 may not be immediately apparent. In this case, we can find a small common factor to start dividing by.

165 and 180 are divisible by 5 (since they end in 5 and 0), so 5 is a common factor that we can divide by to start reducing the fraction.

$$\frac{165\,(\div 5)}{180\,(\div 5)} = \frac{33}{36}$$

We can see that both 33 and 36 are divisible by 3:

$$\frac{33\,(\div 3)}{36\,(\div 3)} = \frac{11}{12}$$

11 and 12 have no common factors other than 1, so this is the simplified fraction.

Name ______________________ Date ______________

Simplifying Fractions

Simplify the fractions.

1. $\frac{4}{8}$ =	2. $\frac{4}{24}$ =	3. $\frac{14}{20}$ =
4. $\frac{9}{12}$ =	5. $\frac{20}{36}$ =	6. $\frac{32}{100}$ =
7. $\frac{20}{44}$ =	8. $\frac{56}{63}$ =	9. $\frac{12}{15}$ =
10. $\frac{63}{70}$ =	11. $\frac{7}{21}$ =	12. $\frac{14}{21}$ =
13. $\frac{45}{225}$ =	14. $\frac{96}{120}$ =	15. $\frac{132}{360}$ =

Name ______________________ Date ______________

Simplifying Fractions

Simplify the fractions.

1. $\frac{8}{12} =$	2. $\frac{40}{56} =$	3. $\frac{4}{14} =$
4. $\frac{42}{54} =$	5. $\frac{4}{6} =$	6. $\frac{6}{12} =$
7. $\frac{27}{45} =$	8. $\frac{28}{100} =$	9. $\frac{24}{33} =$
10. $\frac{10}{25} =$	11. $\frac{36}{44} =$	12. $\frac{12}{36} =$
13. $\frac{64}{100} =$	14. $\frac{80}{112} =$	15. $\frac{126}{315} =$

Name ______________________ Date ______________

Simplifying Fractions

Simplify the fractions.

1. $\frac{9}{45}$ =	2. $\frac{15}{20}$ =	3. $\frac{5}{15}$ =
4. $\frac{3}{18}$ =	5. $\frac{6}{18}$ =	6. $\frac{24}{30}$ =
7. $\frac{10}{15}$ =	8. $\frac{56}{80}$ =	9. $\frac{24}{56}$ =
10. $\frac{4}{36}$ =	11. $\frac{12}{20}$ =	12. $\frac{6}{15}$ =
13. $\frac{72}{126}$ =	14. $\frac{84}{144}$ =	15. $\frac{153}{171}$ =

Name ______________________ Date ______________

Simplifying Fractions

Simplify the fractions.

1. $\frac{5}{10} =$	2. $\frac{16}{20} =$	3. $\frac{18}{50} =$
4. $\frac{24}{27} =$	5. $\frac{8}{24} =$	6. $\frac{28}{35} =$
7. $\frac{12}{32} =$	8. $\frac{9}{18} =$	9. $\frac{20}{30} =$
10. $\frac{15}{21} =$	11. $\frac{18}{45} =$	12. $\frac{14}{50} =$
13. $\frac{64}{128} =$	14. $\frac{75}{360} =$	15. $\frac{125}{325} =$

Name ______________________ Date ______________

Simplifying Fractions

Simplify the fractions.

1. $\frac{8}{28}$ =	2. $\frac{21}{35}$ =	3. $\frac{24}{32}$ =
4. $\frac{32}{40}$ =	5. $\frac{14}{35}$ =	6. $\frac{6}{20}$ =
7. $\frac{8}{40}$ =	8. $\frac{21}{42}$ =	9. $\frac{12}{75}$ =
10. $\frac{16}{36}$ =	11. $\frac{14}{22}$ =	12. $\frac{63}{72}$ =
13. $\frac{72}{168}$ =	14. $\frac{112}{154}$ =	15. $\frac{128}{256}$ =

Converting Improper Fractions to Mixed Numbers

A **proper fraction** is one whose numerator is less than its denominator. These are the fractions that have been covered so far and their value is less than 1. They represent only part of one whole amount.

An **improper fraction** is a fraction whose numerator is greater than its denominator.

$\frac{11}{6}$ is an example of an improper fraction, since 11 is greater than 6.

The value of an improper fraction is greater than 1 and they are often expressed as a mixed number. A **mixed number** consists of an integer and a proper fraction.

$1\frac{5}{6}$ is a mixed number and has the same value as the improper fraction $\frac{11}{6}$.

(The number $1\frac{5}{6}$ is read as "one and five sixths.")

Suppose there were two pizzas, where each pizza was cut into 6 equal slices. If the whole first pizza was eaten (6 out of 6 slices) and 5 slices from the second pizza were also eaten, the total amount eaten could be represented as $1\frac{5}{6}$. The integer 1 represents one whole pizza.

1 (whole pizza) $\quad\quad$ $\frac{5}{6}$

Counting the number of slices eaten, we can see that 11 slices were eaten in all.
Since each slice is $\frac{1}{6}$ of a pizza, $\frac{11}{6}$ (eleven sixths) were eaten, which is the same as $1\frac{5}{6}$.

Fractions are really just division problems (the numerator divided by the denominator), and we can use division to convert an improper fraction to a mixed number.

To convert an improper fraction to a mixed number:

Step 1: Divide the numerator by the denominator, noting the quotient and remainder.

Step 2: Write the mixed number using the results from Step 1.

- A The integer part will be the quotient found in Step 1.
- B The numerator of the fraction part will be the remainder found in Step 1.
- C The denominator will stay the same.

Example: Convert the improper fraction $\frac{7}{3}$ to a mixed number.

Step 1: Divide the numerator by the denominator.

$7 \div 3 = 2$ with a remainder of 1.

$$\begin{array}{r} 2 \text{ R}1 \\ 3\overline{)\,7} \\ -6 \\ \hline 1 \end{array}$$

- A Quotient: 2
- B Remainder: 1
- C Denominator: 3

Step 2: Write the mixed number using the results from Step 1.

- A The integer part will be the quotient (2).
- B The numerator of the fraction part will be the remainder (1).
- C The denominator (3) will stay the same as the original improper fraction.

A

↓

$$\frac{7}{3} = 2\frac{1}{3} \leftarrow \text{B}$$

↑

C

Name ______________________ Date ______________

Improper Fractions to Mixed Numbers

Convert the improper fractions to mixed numbers.

1. $\frac{10}{9}$ ______	2. $\frac{13}{3}$ ______	3. $\frac{6}{5}$ ______
4. $\frac{23}{4}$ ______	5. $\frac{23}{9}$ ______	6. $\frac{29}{7}$ ______
7. $\frac{21}{5}$ ______	8. $\frac{26}{7}$ ______	9. $\frac{11}{4}$ ______
10. $\frac{17}{4}$ ______	11. $\frac{9}{4}$ ______	12. $\frac{35}{6}$ ______
13. $\frac{29}{12}$ ______	14. $\frac{8}{7}$ ______	15. $\frac{23}{10}$ ______

Name ______________________ Date ______________

Improper Fractions to Mixed Numbers

Convert the improper fractions to mixed numbers.

1. $\frac{18}{7}$ ______	2. $\frac{12}{11}$ ______	3. $\frac{10}{7}$ ______
4. $\frac{38}{9}$ ______	5. $\frac{10}{9}$ ______	6. $\frac{43}{6}$ ______
7. $\frac{20}{11}$ ______	8. $\frac{16}{7}$ ______	9. $\frac{9}{8}$ ______
10. $\frac{5}{3}$ ______	11. $\frac{67}{12}$ ______	12. $\frac{11}{3}$ ______
13. $\frac{23}{8}$ ______	14. $\frac{63}{11}$ ______	15. $\frac{19}{7}$ ______

Name ______________________ Date ______________

Improper Fractions to Mixed Numbers

Convert the improper fractions to mixed numbers.

1. $\frac{5}{4}$ ______	2. $\frac{12}{7}$ ______	3. $\frac{16}{9}$ ______
4. $\frac{7}{4}$ ______	5. $\frac{13}{4}$ ______	6. $\frac{11}{10}$ ______
7. $\frac{22}{17}$ ______	8. $\frac{7}{5}$ ______	9. $\frac{25}{12}$ ______
10. $\frac{5}{2}$ ______	11. $\frac{59}{11}$ ______	12. $\frac{33}{10}$ ______
13. $\frac{14}{5}$ ______	14. $\frac{41}{12}$ ______	15. $\frac{15}{7}$ ______

Name ______________________ Date ______________

Improper Fractions to Mixed Numbers

Convert the improper fractions to mixed numbers.

1. $\frac{7}{6}$ ______	2. $\frac{41}{11}$ ______	3. $\frac{9}{5}$ ______
4. $\frac{8}{3}$ ______	5. $\frac{31}{8}$ ______	6. $\frac{24}{5}$ ______
7. $\frac{4}{3}$ ______	8. $\frac{13}{6}$ ______	9. $\frac{11}{2}$ ______
10. $\frac{31}{12}$ ______	11. $\frac{29}{11}$ ______	12. $\frac{46}{9}$ ______
13. $\frac{16}{3}$ ______	14. $\frac{28}{9}$ ______	15. $\frac{23}{5}$ ______

Name ________________________ Date ________________

Improper Fractions to Mixed Numbers

Convert the improper fractions to mixed numbers.

1. $\frac{33}{7}$ ______	2. $\frac{64}{11}$ ______	3. $\frac{37}{9}$ ______
4. $\frac{17}{3}$ ______	5. $\frac{19}{5}$ ______	6. $\frac{25}{6}$ ______
7. $\frac{29}{10}$ ______	8. $\frac{49}{12}$ ______	9. $\frac{9}{2}$ ______
10. $\frac{12}{5}$ ______	11. $\frac{39}{11}$ ______	12. $\frac{11}{7}$ ______
13. $\frac{25}{9}$ ______	14. $\frac{7}{2}$ ______	15. $\frac{19}{4}$ ______

Name ______________________________ Date ____________________

Improper Fractions to Mixed Numbers

Convert the improper fractions to mixed numbers.

1. $\frac{8}{5}$ ______	2. $\frac{18}{5}$ ______	3. $\frac{25}{11}$ ______
4. $\frac{41}{8}$ ______	5. $\frac{3}{2}$ ______	6. $\frac{9}{7}$ ______
7. $\frac{31}{11}$ ______	8. $\frac{27}{7}$ ______	9. $\frac{10}{3}$ ______
10. $\frac{26}{9}$ ______	11. $\frac{52}{11}$ ______	12. $\frac{14}{9}$ ______
13. $\frac{49}{10}$ ______	14. $\frac{17}{9}$ ______	15. $\frac{23}{6}$ ______

Converting Mixed Numbers to Improper Fractions

To convert a mixed number to an improper fraction:

Step 1: Find the numerator of the improper fraction. - Multiply the mixed number's denominator by its whole number part, then add its numerator to that product.

Step 2: Use the numerator found in Step 1 and the denominator of the original mixed number to write the improper fraction.

Example: Convert $2\frac{3}{4}$ to an improper fraction.

Step 1: First multiply the denominator by the whole number part, then add the numerator.

- Denominator: 4
- Whole Number Part: 2
- Numerator: 3

Multiply the denominator by the whole number part
denominator x whole number part = 4 x 2 = 8

Add the numerator
8 + numerator = 8 + 3 = 11

Step 2: Write the improper fraction using the numerator found in the first step (**11**) and the denominator of the original mixed number (**4**).

$$\frac{11}{4}$$

Name ______________________ Date ______________

Improper Fractions to Mixed Numbers

Convert the mixed numbers to improper fractions.

1. $4\frac{1}{6}$ ______	2. $3\frac{2}{7}$ ______	3. $2\frac{7}{8}$ ______
4. $6\frac{1}{4}$ ______	5. $1\frac{7}{9}$ ______	6. $5\frac{5}{8}$ ______
7. $2\frac{2}{5}$ ______	8. $3\frac{9}{11}$ ______	9. $8\frac{3}{5}$ ______
10. $3\frac{7}{8}$ ______	11. $1\frac{2}{3}$ ______	12. $5\frac{7}{9}$ ______
13. $1\frac{4}{5}$ ______	14. $2\frac{7}{10}$ ______	15. $7\frac{4}{5}$ ______

Name ______________________ Date ______________

Improper Fractions to Mixed Numbers

Convert the mixed numbers to improper fractions.

1. $5\frac{6}{11}$ ______	2. $4\frac{7}{9}$ ______	3. $2\frac{9}{10}$ ______
4. $2\frac{1}{6}$ ______	5. $1\frac{9}{11}$ ______	6. $2\frac{1}{7}$ ______
7. $4\frac{1}{2}$ ______	8. $3\frac{3}{7}$ ______	9. $5\frac{4}{5}$ ______
10. $1\frac{2}{7}$ ______	11. $1\frac{1}{4}$ ______	12. $4\frac{3}{5}$ ______
13. $5\frac{3}{4}$ ______	14. $8\frac{2}{3}$ ______	15. $2\frac{1}{4}$ ______

Name ______________________ Date ______________

Improper Fractions to Mixed Numbers

Convert the mixed numbers to improper fractions.

1. $5\frac{1}{8}$ ______	2. $1\frac{8}{9}$ ______	3. $3\frac{1}{3}$ ______
4. $3\frac{7}{9}$ ______	5. $5\frac{7}{10}$ ______	6. $2\frac{1}{5}$ ______
7. $4\frac{3}{4}$ ______	8. $7\frac{1}{5}$ ______	9. $4\frac{4}{7}$ ______
10. $4\frac{2}{3}$ ______	11. $3\frac{4}{5}$ ______	12. $1\frac{3}{8}$ ______
13. $2\frac{5}{8}$ ______	14. $4\frac{5}{11}$ ______	15. $8\frac{9}{10}$ ______

Name ______________________________ Date ______________

Improper Fractions to Mixed Numbers

Convert the mixed numbers to improper fractions.

1. $1\frac{1}{7}$ ______	2. $4\frac{2}{5}$ ______	3. $1\frac{1}{2}$ ______
4. $3\frac{1}{5}$ ______	5. $2\frac{5}{6}$ ______	6. $4\frac{7}{11}$ ______
7. $5\frac{9}{10}$ ______	8. $7\frac{2}{3}$ ______	9. $3\frac{5}{12}$ ______
10. $6\frac{4}{5}$ ______	11. $1\frac{1}{10}$ ______	12. $5\frac{1}{3}$ ______
13. $5\frac{3}{5}$ ______	14. $2\frac{8}{9}$ ______	15. $4\frac{5}{6}$ ______

Name ______________________ Date ______________

Improper Fractions to Mixed Numbers

Convert the mixed numbers to improper fractions.

1. $4\frac{7}{8}$ ______	2. $4\frac{1}{7}$ ______	3. $5\frac{5}{7}$ ______
4. $2\frac{2}{5}$ ______	5. $6\frac{5}{8}$ ______	6. $3\frac{6}{7}$ ______
7. $7\frac{1}{2}$ ______	8. $3\frac{2}{3}$ ______	9. $5\frac{2}{9}$ ______
10. $4\frac{8}{9}$ ______	11. $8\frac{1}{2}$ ______	12. $5\frac{1}{4}$ ______
13. $1\frac{3}{4}$ ______	14. $2\frac{2}{3}$ ______	15. $3\frac{3}{8}$ ______

Fraction Addition and Subtraction

When adding (or subtracting) fractions, the denominators need to be the same.

To add fractions with the same denominators, add the numerators and keep the same denominator as the orginal fractions. The same method works for subtracting. Subtract the numerators and keep the same denominator.

Example: Add and subtract.

a. $\frac{1}{5} + \frac{2}{5} = \frac{1+2}{5} = \frac{3}{5}$

b. $\frac{4}{5} - \frac{3}{5} = \frac{4-3}{5} = \frac{1}{5}$

If the denominators are different, one or both of the fractions must be changed to an equivalent fraction so that both fractions have the same denominator. **The LCM of the denominators can be used as a good common denominator.**

Example: Add the fractions: $\frac{2}{3} + \frac{1}{6}$

Since these have different denominators, equivalent fractions with common denominators need to be used instead. The common denominator should be the LCM of the original denominators (3 and 6).

Step 1: Find a common denominator. The LCM of 3 and 6 is 6, so both fractions should have 6 as the denominator.

Step 2: Find equivalent fractions with common denominators. The only fraction that needs to change is $\frac{2}{3}$, since the second fraction's denominator is already 6.

Multiplying both the numerator and the denominator by 2 would give us an equivalent fraction with 6 as the denominator.

$$\frac{2}{3} = \frac{?}{6} \longrightarrow \frac{2\ (\times 2)}{3\ (\times 2)} = \frac{4}{6}$$

Step 3: Add the equivalent fractions with common denominators.

$$\frac{2}{3} + \frac{1}{6} = \frac{4}{6} + \frac{1}{6} = \frac{4+1}{6} = \frac{5}{6}$$

Use 4/6 instead of 2/3 so that both fractions have the same denominator.

Example: Subtract the fractions: $\frac{7}{10} - \frac{1}{4}$

Since these have different denominators, equivalent fractions with common denominators should be used instead. The common denominator should be the LCM of the original denominators (10 and 4).

Step 1: Find a common denominator. The LCM of 10 and 4 is 20, so both fractions should have 20 as the denominator.

Step 2: Find equivalent fractions with common denominators.

Multiplying both the numerator and the denominator of the **first fraction by 2** gives us an equivalent fraction with **20 as the denominator.**

$$\frac{7}{10} = \frac{?}{20} \longrightarrow \frac{7\ (\times 2)}{10\ (\times 2)} = \frac{14}{20}$$

Multiplying both the numerator and the denominator of the **second fraction by 5** gives us an equivalent fraction with **20 as the denominator**.

$$\frac{1}{4} = \frac{?}{20} \longrightarrow \frac{1\ (\times 5)}{4\ (\times 5)} = \frac{5}{20}$$

Step 3: Subtract the equivalent fractions with common denominators.

$\frac{7}{10} - \frac{1}{4}$ can be rewritten as $\frac{14}{20} - \frac{5}{20}$ with common denominators.

$$\frac{14}{20} - \frac{5}{20} = \frac{14 - 5}{20} = \frac{9}{20}$$

These examples use the LCM of the different denominators to find a common denominator. This is called the Least Common Denominator (or LCD). However, any common multiple of the denominators will work, but using the LCD allows us to work with smaller numbers and less steps will be needed in simplifying the answer.

Name ____________________ Date ____________

Fraction Addition and Subtraction

Add the fractions and simplify. *(Remember to convert improper fractions to mixed numbers.)*

1. $\frac{2}{3} + \frac{1}{5} =$ ____________

2. $\frac{1}{4} + \frac{1}{7} =$ ____________

3. $\frac{5}{9} + \frac{1}{2} =$ ____________

4. $\frac{1}{4} + \frac{2}{3} =$ ____________

5. $\frac{1}{2} + \frac{5}{7} =$ ____________

Subtract the fractions and simplify.

6. $\frac{7}{8} - \frac{1}{4} =$ ____________

7. $\frac{9}{10} - \frac{2}{3} =$ ____________

8. $\frac{1}{2} - \frac{2}{5} =$ ____________

9. $\frac{2}{3} - \frac{4}{7} =$ ____________

10. $\frac{6}{7} - \frac{1}{4} =$ ____________

Name ______________________________ Date ____________________

Fraction Addition and Subtraction

Add the fractions and simplify. *(Remember to convert improper fractions to mixed numbers.)*

1. $\frac{1}{4} + \frac{1}{3} =$ ____________________

2. $\frac{9}{11} + \frac{1}{2} =$ ____________________

3. $\frac{2}{3} + \frac{5}{9} =$ ____________________

4. $\frac{9}{10} + \frac{1}{4} =$ ____________________

5. $\frac{1}{3} + \frac{5}{7} =$ ____________________

Subtract the fractions and simplify.

6. $\frac{3}{4} - \frac{1}{2} =$ ____________________

7. $\frac{3}{4} - \frac{1}{7} =$ ____________________

8. $\frac{1}{2} - \frac{4}{9} =$ ____________________

9. $\frac{7}{8} - \frac{2}{3} =$ ____________________

10. $\frac{2}{3} - \frac{3}{5} =$ ____________________

Name ______________________ Date ______________

Fraction Addition and Subtraction

Add the fractions and simplify. *(Remember to convert improper fractions to mixed numbers.)*

1. $\frac{9}{10} + \frac{3}{4} =$ ____________

2. $\frac{4}{7} + \frac{4}{5} =$ ____________

3. $\frac{1}{2} + \frac{7}{8} =$ ____________

4. $\frac{3}{4} + \frac{2}{3} =$ ____________

5. $\frac{2}{5} + \frac{3}{10} =$ ____________

Subtract the fractions and simplify.

6. $\frac{3}{5} - \frac{1}{3} =$ ____________

7. $\frac{5}{8} - \frac{2}{5} =$ ____________

8. $\frac{8}{9} - \frac{2}{3} =$ ____________

9. $\frac{3}{5} - \frac{1}{10} =$ ____________

10. $\frac{1}{4} - \frac{1}{5} =$ ____________

Name ______________________ Date ______________

Fraction Addition and Subtraction

Add the fractions and simplify. *(Remember to convert improper fractions to mixed numbers.)*

1. $\frac{2}{3} + \frac{1}{6} =$ ______________________

2. $\frac{3}{4} + \frac{1}{9} =$ ______________________

3. $\frac{1}{3} + \frac{3}{4} =$ ______________________

4. $\frac{1}{2} + \frac{9}{10} =$ ______________________

5. $\frac{4}{11} + \frac{2}{3} =$ ______________________

Subtract the fractions and simplify.

6. $\frac{4}{9} - \frac{2}{6} =$ ______________________

7. $\frac{5}{9} - \frac{1}{12} =$ ______________________

8. $\frac{5}{8} - \frac{1}{4} =$ ______________________

9. $\frac{3}{4} - \frac{2}{5} =$ ______________________

10. $\frac{8}{11} - \frac{4}{7} =$ ______________________

Name ______________________ Date ______________

Fraction Addition and Subtraction

Add the fractions and simplify. *(Remember to convert improper fractions to mixed numbers.)*

1. $\frac{2}{5} + \frac{1}{5} =$ ______________

2. $\frac{4}{9} + \frac{2}{3} =$ ______________

3. $\frac{1}{4} + \frac{2}{5} =$ ______________

4. $\frac{1}{2} + \frac{4}{9} =$ ______________

5. $\frac{6}{7} + \frac{1}{4} =$ ______________

Subtract the fractions and simplify.

6. $\frac{2}{3} - \frac{1}{6} =$ ______________

7. $\frac{3}{4} - \frac{4}{9} =$ ______________

8. $\frac{5}{6} - \frac{4}{5} =$ ______________

9. $\frac{7}{9} - \frac{1}{3} =$ ______________

10. $\frac{6}{7} - \frac{3}{5} =$ ______________

Name ______________________ Date ______________

Fraction Addition and Subtraction

Add the fractions and simplify. *(Remember to convert improper fractions to mixed numbers.)*

1. $\frac{3}{5} + \frac{1}{6} =$ ______

2. $\frac{1}{3} + \frac{8}{9} =$ ______

3. $\frac{5}{6} + \frac{2}{5} =$ ______

4. $\frac{2}{9} + \frac{1}{12} =$ ______

5. $\frac{1}{3} + \frac{5}{6} =$ ______

Subtract the fractions and simplify.

6. $\frac{3}{4} - \frac{1}{7} =$ ______

7. $\frac{1}{3} - \frac{1}{9} =$ ______

8. $\frac{3}{5} - \frac{3}{7} =$ ______

9. $\frac{3}{4} - \frac{2}{3} =$ ______

10. $\frac{1}{5} - \frac{1}{6} =$ ______

Name ______________________ Date ____________

Fraction Addition and Subtraction

Add the fractions and simplify. *(Remember to convert improper fractions to mixed numbers.)*

1. $\frac{3}{5} + \frac{1}{2} =$ ______________________

2. $\frac{2}{3} + \frac{1}{9} =$ ______________________

3. $\frac{4}{5} + \frac{3}{7} =$ ______________________

4. $\frac{7}{9} + \frac{2}{3} =$ ______________________

5. $\frac{3}{8} + \frac{1}{6} =$ ______________________

Subtract the fractions and simplify.

6. $\frac{5}{6} - \frac{2}{3} =$ ______________________

7. $\frac{5}{7} - \frac{1}{2} =$ ______________________

8. $\frac{1}{2} - \frac{1}{6} =$ ______________________

9. $\frac{2}{3} - \frac{5}{8} =$ ______________________

10. $\frac{9}{10} - \frac{3}{4} =$ ______________________

Name ______________________ Date ______________

Fraction Addition and Subtraction

Add the fractions and simplify. *(Remember to convert improper fractions to mixed numbers.)*

1. $\frac{5}{9} + \frac{1}{6} =$ ______

2. $\frac{1}{7} + \frac{1}{2} =$ ______

3. $\frac{2}{3} + \frac{6}{7} =$ ______

4. $\frac{1}{2} + \frac{8}{9} =$ ______

5. $\frac{1}{4} + \frac{1}{9} =$ ______

Subtract the fractions and simplify.

6. $\frac{4}{11} - \frac{1}{3} =$ ______

7. $\frac{6}{7} - \frac{4}{5} =$ ______

8. $\frac{3}{4} - \frac{1}{8} =$ ______

9. $\frac{1}{2} - \frac{3}{7} =$ ______

10. $\frac{4}{5} - \frac{3}{4} =$ ______

Adding and Subtracting Mixed Numbers

Adding Mixed Numbers (Method 1)

A mixed number, which consists of a whole number part and a fraction part, can be expressed as the sum of the whole number and the fraction part.

For example, $1\frac{5}{6}$ can be expressed as $1 + \frac{5}{6}$.

To add mixed numbers, we can separate the whole number parts from the fraction parts, add the whole numbers and add the fractions, then put them back together.

Example: Add $1\frac{3}{8} + 2\frac{1}{8}$.

$1\frac{3}{8}$ can be expressed as $1 + \frac{3}{8}$ and $2\frac{1}{8}$ can be expressed as $2 + \frac{1}{8}$.

So the addition problem $1\frac{3}{8} + 2\frac{1}{8}$ can be rewritten as:

$$1 + \frac{3}{8} + 2 + \frac{1}{8}$$

First, add the whole numbers (1 + 2):

$$3 + \frac{3}{8} + \frac{1}{8}$$

Next, add the fractions:

$$3 + \frac{4}{8}$$

Simplify the fraction:

$$3 + \frac{1}{2}$$

$3 + \frac{1}{2}$ can be expressed as the mixed number: $3\frac{1}{2}$

Sometimes the result of adding mixed numbers can end up with a whole number and an improper fraction. In this case, the improper fraction is converted to a mixed number, then the other whole number is added to that. (See the example below.)

Example: Add $2\frac{5}{6} + 3\frac{1}{4}$.

Rewrite each mixed number as the sum of a whole number and a fraction.

$$2 + \frac{5}{6} + 3 + \frac{1}{4}$$

Add the whole numbers (2+3).

$$5 + \frac{5}{6} + \frac{1}{4}$$

Add the fractions. These fractions have different denominators, so they need to be changed to equivalent fractions with the same denominators.

$$5 + \frac{10}{12} + \frac{3}{12}$$

$$5 + \frac{13}{12}$$

Change the improper fraction to a mixed number and rewrite it as the sum of a whole number and a fraction.

$$5 + 1\frac{1}{12}$$

$$5 + 1 + \frac{1}{12}$$

Add the whole numbers together (5+1).

$6 + \frac{1}{12}$ can be expressed as the mixed number: $6\frac{1}{12}$

Adding Mixed Numbers (Method 2)

Another way to add mixed numbers is to convert each one to an improper fraction, add them together, and then convert the answer to a simplified mixed number.

Example: Add $2\frac{5}{6} + 3\frac{1}{4}$.

Rewrite the problem expressing each mixed number as an improper fraction.

$$\frac{17}{6} + \frac{13}{4}$$

Change the fractions to equivalent fractions with common denominators and add them.

$$\frac{34}{12} + \frac{39}{12} = \frac{73}{12}$$

Express as a simplified mixed number.

$$\frac{73}{12} = 6\frac{1}{12}$$

Subtracting Mixed Numbers

To subtract mixed numbers, convert each one to an improper fraction and subtract them.

Subtraction can be done by separating the improper fractions into parts and subtracting the parts, but it can be a more tedious process.

Example: Subtract $5\frac{2}{3} - 4\frac{1}{2}$.

Rewrite the problem expressing each mixed number as an improper fraction.

$$\frac{17}{3} - \frac{9}{2}$$

Change the fractions to equivalent fractions with common denominators and subtract them.

$$\frac{34}{6} - \frac{27}{6} = \frac{7}{6}$$

Express as a simplified mixed number.

$$\frac{7}{6} = 1\frac{1}{6}$$

Name ______________________ Date ______________

Addition and Subtraction with Mixed Numbers

Add the fractions and simplify.

1. $1\frac{1}{3} + 2\frac{1}{3} =$ ____________

2. $5\frac{1}{4} + 3\frac{2}{4} =$ ____________

3. $1\frac{1}{2} + 1\frac{1}{4} =$ ____________

4. $1\frac{1}{5} + 2\frac{1}{2} =$ ____________

5. $1\frac{1}{2} + 1\frac{2}{3} =$ ____________

Subtract the fractions and simplify.

6. $2\frac{1}{5} - 1\frac{2}{5} =$ ____________

7. $3\frac{1}{3} - 2\frac{2}{3} =$ ____________

8. $1\frac{1}{2} - 1\frac{1}{3} =$ ____________

9. $2\frac{1}{3} - 1\frac{1}{4} =$ ____________

10. $3\frac{1}{2} - 2\frac{1}{5} =$ ____________

Name ______________________ Date ______________

Addition and Subtraction with Mixed Numbers

Add the fractions and simplify.

1. $1\frac{3}{5} + 4\frac{1}{5} =$ ______

2. $2\frac{4}{7} + 3\frac{5}{7} =$ ______

3. $2\frac{1}{2} + 2\frac{1}{4} =$ ______

4. $1\frac{1}{8} + 1\frac{1}{2} =$ ______

5. $2\frac{1}{5} + 1\frac{1}{10} =$ ______

Subtract the fractions and simplify.

6. $3\frac{1}{4} - 1\frac{3}{4} =$ ______

7. $2\frac{1}{3} - 1\frac{2}{3} =$ ______

8. $2\frac{1}{4} - 1\frac{1}{8} =$ ______

9. $1\frac{2}{3} - 1\frac{1}{2} =$ ______

10. $2\frac{1}{5} - 1\frac{1}{2} =$ ______

Name ______________________ Date ______________

Addition and Subtraction with Mixed Numbers

Add the fractions and simplify.

1. $1\frac{1}{8} + 3\frac{5}{8} =$ ____________

2. $2\frac{1}{5} + 2\frac{2}{5} =$ ____________

3. $3\frac{1}{2} + 1\frac{2}{3} =$ ____________

4. $1\frac{1}{3} + 1\frac{3}{4} =$ ____________

5. $1\frac{1}{4} + 1\frac{1}{5} =$ ____________

Subtract the fractions and simplify.

6. $3\frac{2}{5} - 1\frac{3}{5} =$ ____________

7. $4\frac{1}{2} - 2\frac{1}{2} =$ ____________

8. $2\frac{1}{8} - 1\frac{1}{4} =$ ____________

9. $1\frac{1}{2} - 1\frac{2}{5} =$ ____________

10. $3\frac{1}{4} - 1\frac{1}{3} =$ ____________

Name ______________________________ Date ____________________

Addition and Subtraction with Mixed Numbers

Add the fractions and simplify.

1. $2\frac{1}{6} + 1\frac{1}{6} =$ ____________________

2. $1\frac{2}{3} + 2\frac{1}{3} =$ ____________________

3. $1\frac{1}{2} + 3\frac{1}{3} =$ ____________________

4. $1\frac{1}{3} + 1\frac{2}{9} =$ ____________________

5. $2\frac{1}{4} + 1\frac{1}{8} =$ ____________________

Subtract the fractions and simplify.

6. $2\frac{3}{8} - 1\frac{1}{8} =$ ____________________

7. $3\frac{1}{7} - 2\frac{3}{7} =$ ____________________

8. $3\frac{1}{2} - 1\frac{2}{8} =$ ____________________

9. $2\frac{1}{5} - 1\frac{1}{4} =$ ____________________

10. $3\frac{3}{4} - 2\frac{1}{2} =$ ____________________

Name ______________________ Date ______________

Addition and Subtraction with Mixed Numbers

Add the fractions and simplify.

1. $1\frac{2}{7} + 3\frac{4}{7} =$ ______

2. $1\frac{3}{5} + 1\frac{4}{5} =$ ______

3. $1\frac{2}{5} + 1\frac{1}{2} =$ ______

4. $2\frac{1}{2} + 1\frac{1}{4} =$ ______

5. $1\frac{1}{2} + 2\frac{1}{3} =$ ______

Subtract the fractions and simplify.

6. $2\frac{1}{5} - 1\frac{3}{5} =$ ______

7. $3\frac{1}{8} - 2\frac{3}{8} =$ ______

8. $1\frac{3}{4} - 1\frac{1}{5} =$ ______

9. $2\frac{1}{3} - 1\frac{1}{2} =$ ______

10. $3\frac{1}{2} - 2\frac{1}{8} =$ ______

Name ______________________ Date ______________

Addition and Subtraction with Mixed Numbers

Add the fractions and simplify.

1. $2\frac{1}{8} + 3\frac{3}{8} =$ ____________________

2. $1\frac{5}{9} + 3\frac{7}{9} =$ ____________________

3. $1\frac{1}{3} + 1\frac{1}{4} =$ ____________________

4. $1\frac{1}{2} + 1\frac{1}{3} =$ ____________________

5. $1\frac{2}{3} + 2\frac{1}{4} =$ ____________________

Subtract the fractions and simplify.

6. $2\frac{1}{6} - 1\frac{5}{6} =$ ____________________

7. $4\frac{2}{3} - 2\frac{1}{3} =$ ____________________

8. $5\frac{1}{2} - 3\frac{1}{4} =$ ____________________

9. $2\frac{3}{5} - 1\frac{1}{2} =$ ____________________

10. $3\frac{1}{2} - 2\frac{1}{10} =$ ____________________

Name ______________________ Date ______________

Addition and Subtraction with Mixed Numbers

Add the fractions and simplify.

1. $4\frac{1}{4} + 2\frac{2}{4} =$ ______________________

2. $2\frac{1}{10} + 3\frac{3}{10} =$ ______________________

3. $1\frac{1}{3} + 2\frac{1}{4} =$ ______________________

4. $2\frac{1}{3} + 1\frac{1}{6} =$ ______________________

5. $1\frac{2}{5} + 1\frac{1}{10} =$ ______________________

Subtract the fractions and simplify.

6. $3\frac{1}{5} - 2\frac{2}{5} =$ ______________________

7. $2\frac{3}{4} - 1\frac{1}{4} =$ ______________________

8. $4\frac{1}{5} - 1\frac{1}{10} =$ ______________________

9. $3\frac{1}{4} - 2\frac{1}{2} =$ ______________________

10. $2\frac{5}{8} - 1\frac{1}{4} =$ ______________________

Fraction Multiplication and Division

Fraction Multiplication

To **multiply fractions**, simply multiply the numerators and multiply the denominators. The denominators do not need to be the same like they do with addition and subtraction.

Example: Multiply: $\frac{3}{5} \times \frac{2}{7}$

Multiply the numerators and multiply the denominators.

$$\frac{3}{5} \times \frac{2}{7} = \frac{3 \times 2}{5 \times 7} = \frac{6}{35}$$

Fraction Division

First let's review some terminology: dividends, divisors and reciprocals.

Dividends and Divisors

The **dividend** is the number that's being divided.
The **divisor** is the number the dividend is divided by.

Reciprocals

The **reciprocal of a fraction** is what you have when the fraction is "flipped" by swapping the numerator and the denominator.

For example, the reciprocal of $\frac{2}{5}$ is $\frac{5}{2}$.

To **divide fractions**, multiply the dividend by the reciprocal of the divisor.

Example: Divide: $\frac{2}{3} \div \frac{3}{4}$

The dividend is $\frac{2}{3}$ and the divisor is $\frac{3}{4}$. The reciprocal of $\frac{3}{4}$ is $\frac{4}{3}$.

Multiply the dividend by the reciprocal of the divisor.

$$\frac{2}{3} \div \frac{3}{4} = \frac{2}{3} \times \frac{4}{3} = \frac{2 \times 4}{3 \times 3} = \frac{8}{9}$$

Dividing a Fraction by a Whole Number

When dividing a fraction by a whole number, we need the reciprocal of that whole number.

To write the reciprocal of a whole number like 3, we can first change it to a fraction, then "flip" the numerator and the denominator.

Remember a fraction is just a division problem, where the numerator is divided by the denominator. A number like 3 can be rewritten as $\frac{3}{1}$. This is just 3 divided by 1, which is 3.

Since 3 is the same as $\frac{3}{1}$, the reciprocal of 3 is $\frac{1}{3}$.

To divide a fraction by a whole number, multiply the fraction by the reciprocal of that whole number. The reciprocal of a whole number will be a fraction with 1 as the numerator and the whole number as the denominator.

Example: Divide: $\frac{1}{2} \div 5$

The dividend is $\frac{1}{2}$ and the divisor is 5. The reciprocal of 5 is $\frac{1}{5}$.

Multiply the dividend by the reciprocal of the divisor.

$$\frac{1}{2} \div 5 = \frac{1}{2} \times \frac{1}{5} = \frac{1 \times 1}{2 \times 5} = \frac{1}{10}$$

Name ______________________ Date ______________

Fraction Muliplication and Division

Multiply the fractions and simplify.

1. $\frac{6}{7} \times \frac{2}{3} =$ ______________

2. $\frac{2}{7} \times \frac{7}{12} =$ ______________

3. $\frac{3}{4} \times \frac{7}{9} =$ ______________

4. $\frac{2}{3} \times \frac{7}{10} =$ ______________

5. $\frac{1}{2} \times \frac{7}{11} =$ ______________

Divide the fractions and simplify.

6. $\frac{1}{6} \div \frac{1}{2} =$ ______________

7. $\frac{4}{5} \div \frac{3}{8} =$ ______________

8. $\frac{1}{2} \div \frac{8}{11} =$ ______________

9. $\frac{2}{5} \div \frac{1}{5} =$ ______________

10. $\frac{3}{8} \div 2 =$ ______________

Name ______________________ Date ______________

Fraction Muliplication and Division

Multiply the fractions and simplify.

1. $\frac{8}{11} \times \frac{3}{4} =$ ______________

2. $\frac{2}{3} \times \frac{5}{6} =$ ______________

3. $\frac{1}{2} \times \frac{3}{4} =$ ______________

4. $\frac{1}{2} \times \frac{2}{3} =$ ______________

5. $\frac{4}{5} \times \frac{7}{10} =$ ______________

Divide the fractions and simplify.

6. $\frac{3}{4} \div \frac{5}{9} =$ ______________

7. $\frac{3}{10} \div \frac{9}{10} =$ ______________

8. $\frac{4}{11} \div 4 =$ ______________

9. $\frac{3}{5} \div \frac{4}{5} =$ ______________

10. $\frac{2}{3} \div \frac{3}{8} =$ ______________

Name ______________________ Date ______________

Fraction Muliplication and Division

Multiply the fractions and simplify.

1. $\frac{5}{12} \times \frac{1}{2} =$ ______________

2. $\frac{3}{7} \times \frac{1}{6} =$ ______________

3. $\frac{1}{3} \times \frac{1}{2} =$ ______________

4. $\frac{6}{11} \times \frac{1}{4} =$ ______________

5. $\frac{4}{5} \times \frac{4}{5} =$ ______________

Divide the fractions and simplify.

6. $\frac{3}{10} \div \frac{2}{3} =$ ______________

7. $\frac{5}{7} \div \frac{5}{6} =$ ______________

8. $\frac{5}{8} \div 5 =$ ______________

9. $\frac{7}{9} \div \frac{5}{12} =$ ______________

10. $\frac{1}{2} \div \frac{3}{4} =$ ______________

Name ______________________________ Date ________________

Fraction Muliplication and Division

Multiply the fractions and simplify.

1. $\frac{9}{10} \times \frac{1}{3} =$ ________________

2. $\frac{5}{6} \times \frac{9}{10} =$ ________________

3. $\frac{3}{7} \times \frac{7}{12} =$ ________________

4. $\frac{5}{6} \times \frac{3}{4} =$ ________________

5. $\frac{2}{3} \times \frac{1}{5} =$ ________________

Divide the fractions and simplify.

6. $\frac{7}{11} \div \frac{5}{11} =$ ________________

7. $\frac{5}{9} \div 4 =$ ________________

8. $\frac{2}{3} \div \frac{3}{4} =$ ________________

9. $\frac{3}{4} \div \frac{1}{2} =$ ________________

10. $\frac{1}{6} \div \frac{2}{3} =$ ________________

Name ______________________ Date ______________

Fraction Muliplication and Division

Multiply the fractions and simplify.

1. $\frac{5}{12} \times \frac{6}{7} =$ ______

2. $\frac{3}{4} \times \frac{1}{5} =$ ______

3. $\frac{4}{7} \times \frac{1}{2} =$ ______

4. $\frac{3}{5} \times \frac{2}{3} =$ ______

5. $\frac{2}{7} \times \frac{3}{4} =$ ______

Divide the fractions and simplify.

6. $\frac{3}{5} \div \frac{9}{10} =$ ______

7. $\frac{4}{5} \div \frac{3}{7} =$ ______

8. $\frac{6}{11} \div \frac{1}{2} =$ ______

9. $\frac{5}{9} \div 3 =$ ______

10. $\frac{2}{3} \div \frac{4}{7} =$ ______

Name ________________________ Date ____________

Fraction Muliplication and Division

Multiply the fractions and simplify.

1. $\frac{3}{4} \times \frac{6}{11} =$ ____________

2. $\frac{1}{5} \times \frac{4}{5} =$ ____________

3. $\frac{5}{8} \times \frac{2}{3} =$ ____________

4. $\frac{2}{9} \times \frac{3}{4} =$ ____________

5. $\frac{8}{11} \times \frac{1}{2} =$ ____________

Divide the fractions and simplify.

6. $\frac{4}{5} \div 2 =$ ____________

7. $\frac{3}{4} \div \frac{2}{3} =$ ____________

8. $\frac{1}{3} \div \frac{1}{5} =$ ____________

9. $\frac{5}{12} \div \frac{1}{2} =$ ____________

10. $\frac{2}{3} \div \frac{5}{12} =$ ____________

Name ______________________ Date ______________

Fraction Muliplication and Division

Multiply the fractions and simplify.

1. $\frac{4}{9} \times \frac{9}{10} =$ ______________________

2. $\frac{1}{4} \times \frac{1}{4} =$ ______________________

3. $\frac{1}{6} \times \frac{3}{4} =$ ______________________

4. $\frac{7}{12} \times \frac{3}{4} =$ ______________________

5. $\frac{1}{2} \times \frac{3}{5} =$ ______________________

Divide the fractions and simplify.

6. $\frac{9}{10} \div \frac{1}{3} =$ ______________________

7. $\frac{1}{4} \div \frac{2}{3} =$ ______________________

8. $\frac{6}{7} \div 3 =$ ______________________

9. $\frac{1}{2} \div \frac{9}{10} =$ ______________________

10. $\frac{2}{7} \div \frac{2}{5} =$ ______________________

Name ________________________________ Date ____________________

Fraction Muliplication and Division

Multiply the fractions and simplify.

1. $\frac{7}{10} \times \frac{6}{7} =$ ____________________

2. $\frac{7}{8} \times \frac{4}{5} =$ ____________________

3. $\frac{4}{5} \times \frac{2}{3} =$ ____________________

4. $\frac{2}{9} \times \frac{1}{4} =$ ____________________

5. $\frac{2}{3} \times \frac{9}{11} =$ ____________________

Divide the fractions and simplify.

6. $\frac{4}{7} \div \frac{1}{2} =$ ____________________

7. $\frac{5}{6} \div \frac{1}{3} =$ ____________________

8. $\frac{1}{2} \div 2 =$ ____________________

9. $\frac{7}{9} \div \frac{7}{8} =$ ____________________

10. $\frac{3}{8} \div \frac{3}{10} =$ ____________________

Name ____________________ Date ____________

Fraction Muliplication and Division

Multiply the fractions and simplify.

1. $\frac{7}{8} \times \frac{8}{9} =$ ____________

2. $\frac{2}{3} \times \frac{6}{11} =$ ____________

3. $\frac{1}{4} \times \frac{1}{3} =$ ____________

4. $\frac{1}{2} \times \frac{5}{9} =$ ____________

5. $\frac{5}{9} \times \frac{3}{5} =$ ____________

Divide the fractions and simplify.

6. $\frac{3}{5} \div \frac{5}{6} =$ ____________

7. $\frac{1}{2} \div \frac{6}{7} =$ ____________

8. $\frac{7}{10} \div \frac{1}{2} =$ ____________

9. $\frac{2}{3} \div 4 =$ ____________

10. $\frac{5}{6} \div \frac{3}{4} =$ ____________

Name ______________________ Date ______________

Fraction Muliplication and Division

Multiply the fractions and simplify.

1. $\frac{1}{2} \times \frac{5}{6} =$ ______________________

2. $\frac{5}{7} \times \frac{1}{3} =$ ______________________

3. $\frac{3}{8} \times \frac{8}{9} =$ ______________________

4. $\frac{3}{4} \times \frac{4}{5} =$ ______________________

5. $\frac{1}{6} \times \frac{9}{11} =$ ______________________

Divide the fractions and simplify.

6. $\frac{1}{3} \div 3 =$ ______________________

7. $\frac{1}{2} \div \frac{3}{6} =$ ______________________

8. $\frac{5}{6} \div \frac{1}{2} =$ ______________________

9. $\frac{2}{3} \div \frac{6}{7} =$ ______________________

10. $\frac{7}{10} \div \frac{2}{3} =$ ______________________

Solutions

Solutions to Problems

Divisibility Rules

Circle the numbers (2 through 10) that divide evenly into the given numbers.

1. 88

2 3 **4** 5 6 7 **8** 9 10

2. 56

2 3 **4** 5 6 **7** **8** 9 10

3. 360

2 **3** **4** **5** **6** 7 **8** **9** **10**

4. 786

2 **3** 4 5 **6** 7 8 9 10

5. 230

2 3 4 **5** 6 7 8 9 **10**

6. 225

2 **3** 4 **5** 6 7 8 **9** 10

7. 2,394

2 **3** 4 5 **6** **7** 8 **9** 10

8. 4,130

2 3 4 **5** 6 **7** 8 9 **10**

9. 7,072

2 3 **4** 5 6 7 **8** 9 10

10. 12,168

2 **3** **4** 5 **6** 7 **8** **9** 10

PAGE 3

Divisibility Rules

Circle the numbers (2 through 10) that divide evenly into the given numbers.

1. 45

2 **3** 4 **5** 6 7 8 **9** 10

2. 54

2 **3** 4 5 **6** 7 8 **9** 10

3. 174

2 **3** 4 5 **6** 7 8 9 10

4. 816

2 **3** **4** 5 **6** 7 **8** 9 10

5. 728

2 3 **4** 5 6 **7** **8** 9 10

6. 645

2 **3** 4 **5** 6 7 8 9 10

7. 5,635

2 3 4 **5** 6 **7** 8 9 10

8. 3,610

2 3 4 **5** 6 7 8 9 **10**

9. 9,846

2 **3** 4 5 **6** 7 8 **9** 10

10. 11,944

2 3 **4** 5 6 7 **8** 9 10

PAGE 4

Squares and Square Roots

Fill in the chart with the squares for numbers 1 through 12 to use with the problems below.

1	2	3	4	5	6	7	8	9	10	11	12
1	4	9	16	25	36	49	64	81	100	121	144

Fill in the blanks with the two consecutive perfect squares that the given numbers are between and write the **square roots** of the lower ones in the grey boxes. Use the chart for reference.

$\sqrt{16} = 4$

1. [4] 16 < 20 < 25
2. [6] 36 < 40 < 49
3. [6] 36 < 48 < 49
4. [7] 49 < 52 < 64
5. [8] 64 < 65 < 81
6. [8] 64 < 80 < 81
7. [9] 81 < 95 < 100
8. [11] 121 < 132 < 144

PAGE 7

Find the Factors

List the factors of the numbers below. If a given number is prime, write prime.

1. 8 — 1, 2, 4, 8
2. 15 — 1, 3, 5, 15
3. 17 — 1, 17 prime
4. 28 — 1, 2, 4, 7, 14, 28
5. 32 — 1, 2, 4, 8, 16, 32
6. 40 — 1, 2, 4, 5, 8, 10, 20, 40
7. 65 — 1, 5, 13, 65
8. 100 — 1, 2, 4, 5, 10, 20, 25, 50, 100
9. 124 — 1, 2, 4, 31, 62, 124
10. 130 — 1, 2, 5, 10, 13, 26, 65, 130

PAGE 9

Find the Factors

List the factors of the numbers below. If a given number is prime, write prime.

1. 9 — 1, 3, 9
2. 24 — 1, 2, 3, 4, 6, 8, 12, 24
3. 36 — 1, 2, 3, 4, 6, 9, 12, 18, 36
4. 45 — 1, 3, 5, 9, 15, 45
5. 51 — 1, 3, 17, 51
6. 88 — 1, 2, 4, 8, 11, 22, 44, 88
7. 85 — 1, 5, 17, 85
8. 113 — 1, 113 prime
9. 115 — 1, 5, 23, 115
10. 135 — 1, 3, 5, 9, 15, 27, 45, 135

PAGE 10

Find the Factors

List the factors of the numbers below. If a given number is prime, write prime.

1. 10 — 1, 2, 5, 10
2. 25 — 1, 5, 25
3. 42 — 1, 2, 3, 6, 7, 14, 21, 42
4. 70 — 1, 2, 5, 7, 10, 14, 35, 70
5. 84 — 1, 2, 3, 4, 6, 7, 12, 14, 21, 28, 42, 84
6. 91 — 1, 7, 13, 91
7. 110 — 1, 2, 5, 10, 11, 22, 55, 110
8. 116 — 1, 2, 4, 29, 58, 116
9. 127 — 1, 127 prime
10. 138 — 1, 2, 3, 6, 23, 46, 69, 138

PAGE 11

GCF (Greatest Common Factor)

Find the GCF.

1. 6, 8 — GCF: 2
2. 10, 55 — GCF: 5
3. 13, 26 — GCF: 13
4. 15, 65 — GCF: 5
5. 16, 22 — GCF: 2
6. 24, 39 — GCF: 3
7. 30, 45 — GCF: 15
8. 49, 56 — GCF: 7
9. 8, 36, 56 — GCF: 4
10. 22, 55, 88 — GCF: 11

PAGE 13

GCF (Greatest Common Factor)

Find the GCF.

1. 6, 24 — GCF: 6
2. 14, 28 — GCF: 14
3. 15, 45 — GCF: 15
4. 21, 28 — GCF: 7
5. 24, 36 — GCF: 12
6. 30, 42 — GCF: 6
7. 50, 125 — GCF: 25
8. 56, 120 — GCF: 8
9. 3, 5, 11 — GCF: 1
10. 12, 36, 72 — GCF: 12

PAGE 14

GCF (Greatest Common Factor)

Find the GCF.

1. 8, 12 — GCF: 4
2. 15, 33 — GCF: 3
3. 21, 35 — GCF: 7
4. 24, 90 — GCF: 6
5. 34, 39 — GCF: 1
6. 35, 40 — GCF: 5
7. 48, 96 — GCF: 48
8. 70, 140 — GCF: 70
9. 12, 24, 45 — GCF: 3
10. 30, 48, 54 — GCF: 6

PAGE 15

GCF (Greatest Common Factor)

Find the GCF.

1. 12, 15 — GCF: 3
2. 13, 39 — GCF: 13
3. 16, 18 — GCF: 2
4. 21, 27 — GCF: 3
5. 22, 121 — GCF: 11
6. 26, 34 — GCF: 2
7. 45, 55 — GCF: 5
8. 36, 54 — GCF: 18
9. 15, 25, 55 — GCF: 5
10. 24, 36, 60 — GCF: 12

PAGE 16

LCM (Least Common Multiple)

Find the LCM.

1. 2, 4 — LCM: 4
2. 4, 6 — LCM: 12
3. 5, 11 — LCM: 55
4. 6, 8 — LCM: 24
5. 6, 10 — LCM: 30
6. 5, 12 — LCM: 60
7. 9, 15 — LCM: 45
8. 9, 12 — LCM: 36
9. 12, 30 — LCM: 60
10. 16, 24 — LCM: 48

PAGE 19

LCM (Least Common Multiple)

Find the LCM.

1. 2, 3 — LCM: 6
2. 3, 7 — LCM: 21
3. 4, 10 — LCM: 20
4. 6, 12 — LCM: 12
5. 6, 20 — LCM: 60
6. 8, 9 — LCM: 72
7. 8, 11 — LCM: 88
8. 10, 25 — LCM: 50
9. 12, 15 — LCM: 60
10. 15, 45 — LCM: 45

PAGE 20

LCM (Least Common Multiple)

Find the LCM.

1. 2, 5 — LCM: 10
2. 3, 25 — LCM: 75
3. 4, 50 — LCM: 100
4. 5, 9 — LCM: 45
5. 9, 21 — LCM: 63
6. 6, 15 — LCM: 30
7. 10, 15 — LCM: 30
8. 10, 45 — LCM: 90
9. 12, 20 — LCM: 60
10. 15, 25 — LCM: 75

PAGE 21

LCM (Least Common Multiple)

Find the LCM.

1. 2, 7 — LCM: 14
2. 4, 8 — LCM: 8
3. 5, 12 — LCM: 60
4. 6, 9 — LCM: 18
5. 8, 12 — LCM: 24
6. 10, 12 — LCM: 60
7. 12, 16 — LCM: 48
8. 8, 20 — LCM: 40
9. 15, 30 — LCM: 30
10. 20, 25 — LCM: 100

PAGE 22

Equivalent Fractions

Fill in each blank with the missing numerator or denominator to make the fractions equivalent.

1. $\frac{2}{9} = \frac{18}{81}$
2. $\frac{5}{6} = \frac{20}{24}$
3. $\frac{7}{10} = \frac{49}{70}$
4. $\frac{1}{2} = \frac{3}{6}$
5. $\frac{3}{9} = \frac{21}{63}$
6. $\frac{1}{8} = \frac{2}{16}$
7. $\frac{9}{25} = \frac{36}{100}$
8. $\frac{2}{6} = \frac{4}{12}$
9. $\frac{1}{5} = \frac{4}{20}$
10. $\frac{2}{8} = \frac{10}{40}$
11. $\frac{4}{10} = \frac{8}{20}$
12. $\frac{5}{8} = \frac{25}{40}$
13. $\frac{5}{7} = \frac{35}{49}$
14. $\frac{6}{12} = \frac{12}{24}$
15. $\frac{3}{4} = \frac{6}{8}$
16. $\frac{7}{12} = \frac{21}{36}$
17. $\frac{1}{9} = \frac{9}{81}$
18. $\frac{9}{10} = \frac{45}{50}$
19. $\frac{7}{11} = \frac{28}{44}$
20. $\frac{9}{12} = \frac{18}{24}$
21. $\frac{8}{11} = \frac{48}{66}$
22. $\frac{6}{8} = \frac{60}{80}$
23. $\frac{5}{100} = \frac{10}{200}$
24. $\frac{4}{5} = \frac{40}{50}$
25. $\frac{3}{12} = \frac{21}{84}$
26. $\frac{4}{12} = \frac{12}{36}$
27. $\frac{8}{9} = \frac{56}{63}$
28. $\frac{3}{11} = \frac{24}{88}$
29. $\frac{11}{12} = \frac{55}{60}$
30. $\frac{8}{12} = \frac{32}{48}$

PAGE 25

Equivalent Fractions

Fill in each blank with the missing numerator or denominator to make the fractions equivalent.

1. $\frac{3}{7} = \frac{12}{28}$
2. $\frac{2}{3} = \frac{14}{21}$
3. $\frac{4}{6} = \frac{28}{42}$
4. $\frac{1}{3} = \frac{5}{15}$
5. $\frac{7}{8} = \frac{42}{48}$
6. $\frac{2}{4} = \frac{20}{40}$
7. $\frac{10}{11} = \frac{50}{55}$
8. $\frac{1}{2} = \frac{4}{8}$
9. $\frac{6}{50} = \frac{12}{100}$
10. $\frac{2}{8} = \frac{4}{16}$
11. $\frac{8}{11} = \frac{64}{88}$
12. $\frac{1}{5} = \frac{8}{40}$
13. $\frac{8}{25} = \frac{16}{50}$
14. $\frac{7}{9} = \frac{28}{36}$
15. $\frac{2}{12} = \frac{6}{36}$
16. $\frac{9}{12} = \frac{36}{48}$
17. $\frac{3}{11} = \frac{15}{55}$
18. $\frac{3}{10} = \frac{6}{20}$
19. $\frac{4}{100} = \frac{16}{400}$
20. $\frac{6}{12} = \frac{30}{60}$
21. $\frac{3}{8} = \frac{21}{56}$
22. $\frac{7}{12} = \frac{14}{24}$
23. $\frac{4}{7} = \frac{8}{14}$
24. $\frac{5}{8} = \frac{15}{24}$
25. $\frac{2}{11} = \frac{4}{22}$
26. $\frac{10}{12} = \frac{100}{120}$
27. $\frac{4}{9} = \frac{32}{72}$
28. $\frac{6}{9} = \frac{30}{45}$
29. $\frac{8}{9} = \frac{72}{81}$
30. $\frac{5}{10} = \frac{50}{100}$

PAGE 26

Equivalent Fractions

Fill in each blank with the missing numerator or denominator to make the fractions equivalent.

1. $\frac{6}{10} = \frac{24}{40}$	2. $\frac{5}{6} = \frac{35}{42}$	3. $\frac{4}{11} = \frac{28}{77}$
4. $\frac{5}{25} = \frac{15}{75}$	5. $\frac{1}{4} = \frac{2}{8}$	6. $\frac{2}{9} = \frac{8}{36}$
7. $\frac{1}{2} = \frac{4}{8}$	8. $\frac{4}{9} = \frac{20}{45}$	9. $\frac{2}{7} = \frac{6}{21}$
10. $\frac{5}{8} = \frac{30}{48}$	11. $\frac{3}{5} = \frac{6}{10}$	12. $\frac{10}{12} = \frac{50}{60}$
13. $\frac{2}{11} = \frac{10}{55}$	14. $\frac{8}{10} = \frac{48}{60}$	15. $\frac{3}{4} = \frac{30}{40}$
16. $\frac{4}{12} = \frac{20}{60}$	17. $\frac{8}{9} = \frac{32}{36}$	18. $\frac{5}{12} = \frac{60}{144}$
19. $\frac{1}{3} = \frac{8}{24}$	20. $\frac{11}{12} = \frac{33}{36}$	21. $\frac{5}{7} = \frac{10}{14}$
22. $\frac{8}{25} = \frac{16}{50}$	23. $\frac{2}{5} = \frac{12}{30}$	24. $\frac{7}{8} = \frac{77}{88}$
25. $\frac{3}{6} = \frac{15}{30}$	26. $\frac{2}{4} = \frac{6}{12}$	27. $\frac{3}{8} = \frac{12}{32}$
28. $\frac{6}{7} = \frac{48}{56}$	29. $\frac{2}{3} = \frac{4}{6}$	30. $\frac{7}{9} = \frac{35}{45}$

PAGE 27

Equivalent Fractions

Fill in each blank with the missing numerator or denominator to make the fractions equivalent.

1. $\frac{4}{8} = \frac{32}{64}$	2. $\frac{1}{10} = \frac{10}{100}$	3. $\frac{3}{7} = \frac{24}{56}$
4. $\frac{1}{7} = \frac{8}{56}$	5. $\frac{2}{3} = \frac{6}{9}$	6. $\frac{1}{4} = \frac{4}{16}$
7. $\frac{2}{5} = \frac{14}{35}$	8. $\frac{4}{7} = \frac{20}{35}$	9. $\frac{5}{9} = \frac{40}{72}$
10. $\frac{7}{10} = \frac{42}{60}$	11. $\frac{3}{5} = \frac{33}{55}$	12. $\frac{7}{11} = \frac{49}{77}$
13. $\frac{4}{10} = \frac{24}{60}$	14. $\frac{3}{8} = \frac{30}{80}$	15. $\frac{4}{8} = \frac{16}{32}$
16. $\frac{5}{11} = \frac{25}{55}$	17. $\frac{5}{6} = \frac{15}{18}$	18. $\frac{3}{5} = \frac{27}{45}$
19. $\frac{6}{8} = \frac{12}{16}$	20. $\frac{5}{8} = \frac{20}{32}$	21. $\frac{2}{6} = \frac{10}{30}$
22. $\frac{4}{25} = \frac{16}{100}$	23. $\frac{3}{9} = \frac{21}{63}$	24. $\frac{11}{12} = \frac{77}{84}$
25. $\frac{8}{9} = \frac{16}{18}$	26. $\frac{9}{12} = \frac{27}{36}$	27. $\frac{6}{12} = \frac{24}{48}$
28. $\frac{7}{12} = \frac{28}{48}$	29. $\frac{8}{11} = \frac{24}{33}$	30. $\frac{9}{10} = \frac{72}{80}$

PAGE 28

Comparing Fractions

Compare the fractions and fill in the blanks with < or >.

1. $\frac{5}{6} > \frac{2}{11}$	2. $\frac{2}{3} > \frac{4}{8}$	3. $\frac{3}{6} < \frac{6}{10}$
4. $\frac{8}{11} < \frac{6}{7}$	5. $\frac{7}{9} > \frac{2}{3}$	6. $\frac{6}{8} > \frac{3}{7}$
7. $\frac{7}{9} > \frac{5}{8}$	8. $\frac{8}{9} > \frac{5}{7}$	9. $\frac{3}{9} > \frac{2}{8}$
10. $\frac{4}{6} > \frac{1}{4}$	11. $\frac{6}{12} < \frac{5}{8}$	12. $\frac{2}{6} < \frac{5}{7}$
13. $\frac{3}{5} > \frac{6}{11}$	14. $\frac{5}{9} < \frac{5}{6}$	15. $\frac{5}{8} > \frac{3}{12}$
16. $\frac{5}{12} < \frac{2}{4}$	17. $\frac{6}{10} < \frac{6}{9}$	18. $\frac{4}{5} < \frac{8}{9}$
19. $\frac{1}{4} < \frac{4}{5}$	20. $\frac{7}{11} > \frac{2}{6}$	21. $\frac{4}{10} < \frac{5}{8}$
22. $\frac{3}{8} < \frac{7}{10}$	23. $\frac{2}{3} < \frac{9}{12}$	24. $\frac{4}{7} > \frac{3}{9}$
25. $\frac{3}{10} > \frac{1}{11}$	26. $\frac{9}{12} < \frac{8}{10}$	27. $\frac{7}{9} < \frac{5}{6}$
28. $\frac{1}{4} < \frac{4}{7}$	29. $\frac{4}{5} > \frac{3}{8}$	30. $\frac{4}{7} > \frac{4}{11}$

PAGE 31

Comparing Fractions

Compare the fractions and fill in the blanks with < or >.

1. $\frac{1}{3} < \frac{9}{12}$	2. $\frac{7}{10} > \frac{2}{9}$	3. $\frac{5}{12} < \frac{5}{6}$
4. $\frac{5}{7} > \frac{1}{10}$	5. $\frac{3}{8} > \frac{2}{6}$	6. $\frac{3}{7} > \frac{2}{5}$
7. $\frac{3}{10} < \frac{3}{6}$	8. $\frac{7}{9} < \frac{9}{10}$	9. $\frac{3}{4} > \frac{7}{11}$
10. $\frac{2}{3} > \frac{7}{12}$	11. $\frac{5}{8} < \frac{3}{4}$	12. $\frac{2}{7} < \frac{7}{12}$
13. $\frac{3}{9} < \frac{5}{7}$	14. $\frac{8}{10} > \frac{2}{12}$	15. $\frac{9}{12} < \frac{7}{9}$
16. $\frac{1}{4} > \frac{2}{9}$	17. $\frac{5}{9} < \frac{7}{8}$	18. $\frac{7}{8} > \frac{3}{5}$
19. $\frac{1}{8} < \frac{4}{7}$	20. $\frac{7}{12} < \frac{7}{11}$	21. $\frac{2}{5} < \frac{7}{10}$
22. $\frac{3}{8} > \frac{2}{7}$	23. $\frac{2}{5} > \frac{3}{8}$	24. $\frac{2}{3} < \frac{8}{9}$
25. $\frac{2}{12} < \frac{5}{6}$	26. $\frac{3}{5} < \frac{5}{7}$	27. $\frac{4}{8} < \frac{5}{6}$
28. $\frac{6}{7} > \frac{4}{5}$	29. $\frac{3}{4} < \frac{5}{6}$	30. $\frac{2}{9} < \frac{6}{10}$

PAGE 32

Comparing Fractions

Compare the fractions and fill in the blanks with < or >.

1. $\frac{7}{10} > \frac{6}{11}$	2. $\frac{6}{8} > \frac{2}{3}$	3. $\frac{1}{5} < \frac{3}{7}$
4. $\frac{7}{9} > \frac{7}{12}$	5. $\frac{2}{7} < \frac{2}{4}$	6. $\frac{8}{10} > \frac{5}{12}$
7. $\frac{4}{5} < \frac{6}{7}$	8. $\frac{9}{12} > \frac{8}{11}$	9. $\frac{3}{4} > \frac{5}{8}$
10. $\frac{1}{6} < \frac{1}{5}$	11. $\frac{4}{5} > \frac{5}{8}$	12. $\frac{3}{7} < \frac{4}{9}$
13. $\frac{8}{11} < \frac{8}{9}$	14. $\frac{8}{10} > \frac{1}{2}$	15. $\frac{4}{7} > \frac{3}{12}$
16. $\frac{5}{9} > \frac{3}{12}$	17. $\frac{5}{6} < \frac{7}{8}$	18. $\frac{5}{12} < \frac{6}{8}$
19. $\frac{6}{8} > \frac{5}{7}$	20. $\frac{9}{11} > \frac{3}{5}$	21. $\frac{7}{9} < \frac{8}{10}$
22. $\frac{4}{6} < \frac{8}{9}$	23. $\frac{3}{8} > \frac{2}{6}$	24. $\frac{3}{4} > \frac{6}{11}$
25. $\frac{7}{12} < \frac{9}{11}$	26. $\frac{6}{7} > \frac{2}{8}$	27. $\frac{9}{10} > \frac{6}{7}$
28. $\frac{3}{6} > \frac{1}{5}$	29. $\frac{1}{4} < \frac{2}{5}$	30. $\frac{3}{6} < \frac{4}{5}$

PAGE 33

Comparing Fractions

Compare the fractions and fill in the blanks with < or >.

1. $\frac{3}{5} > \frac{2}{4}$	2. $\frac{6}{8} > \frac{6}{10}$	3. $\frac{3}{12} < \frac{4}{8}$
4. $\frac{9}{11} > \frac{7}{12}$	5. $\frac{3}{7} > \frac{2}{9}$	6. $\frac{7}{8} > \frac{5}{12}$
7. $\frac{2}{6} < \frac{8}{9}$	8. $\frac{9}{12} < \frac{8}{9}$	9. $\frac{1}{5} < \frac{7}{9}$
10. $\frac{5}{9} < \frac{4}{7}$	11. $\frac{1}{4} > \frac{1}{5}$	12. $\frac{2}{6} > \frac{1}{4}$
13. $\frac{4}{5} > \frac{7}{12}$	14. $\frac{1}{3} < \frac{3}{8}$	15. $\frac{3}{7} > \frac{3}{9}$
16. $\frac{3}{8} < \frac{5}{7}$	17. $\frac{4}{5} < \frac{9}{10}$	18. $\frac{3}{5} > \frac{4}{7}$
19. $\frac{9}{10} > \frac{3}{4}$	20. $\frac{7}{9} > \frac{1}{6}$	21. $\frac{2}{3} < \frac{8}{10}$
22. $\frac{2}{8} < \frac{6}{7}$	23. $\frac{3}{11} < \frac{6}{9}$	24. $\frac{2}{9} < \frac{5}{11}$
25. $\frac{8}{12} > \frac{3}{8}$	26. $\frac{6}{9} < \frac{8}{11}$	27. $\frac{5}{6} < \frac{8}{9}$
28. $\frac{8}{11} < \frac{8}{10}$	29. $\frac{8}{10} > \frac{8}{12}$	30. $\frac{4}{10} > \frac{2}{8}$

PAGE 34

Simplifying Fractions

Simplify the fractions.

1. $\frac{4}{8} = \frac{1}{2}$	2. $\frac{4}{24} = \frac{1}{6}$	3. $\frac{14}{20} = \frac{7}{10}$
4. $\frac{9}{12} = \frac{3}{4}$	5. $\frac{20}{36} = \frac{5}{9}$	6. $\frac{32}{100} = \frac{8}{25}$
7. $\frac{20}{44} = \frac{5}{11}$	8. $\frac{56}{63} = \frac{8}{9}$	9. $\frac{12}{15} = \frac{4}{5}$
10. $\frac{63}{70} = \frac{9}{10}$	11. $\frac{7}{21} = \frac{1}{3}$	12. $\frac{14}{21} = \frac{2}{3}$
13. $\frac{45}{225} = \frac{1}{5}$	14. $\frac{96}{120} = \frac{4}{5}$	15. $\frac{132}{360} = \frac{11}{30}$

PAGE 36

Simplifying Fractions

Simplify the fractions.

1. $\frac{8}{12} = \frac{2}{3}$	2. $\frac{40}{56} = \frac{5}{7}$	3. $\frac{4}{14} = \frac{2}{7}$
4. $\frac{42}{54} = \frac{7}{9}$	5. $\frac{4}{6} = \frac{2}{3}$	6. $\frac{6}{12} = \frac{1}{2}$
7. $\frac{27}{45} = \frac{3}{5}$	8. $\frac{28}{100} = \frac{7}{25}$	9. $\frac{24}{33} = \frac{8}{11}$
10. $\frac{10}{25} = \frac{2}{5}$	11. $\frac{36}{44} = \frac{9}{11}$	12. $\frac{12}{36} = \frac{1}{3}$
13. $\frac{64}{100} = \frac{16}{25}$	14. $\frac{80}{112} = \frac{5}{7}$	15. $\frac{126}{315} = \frac{2}{5}$

PAGE 37

Simplifying Fractions

Simplify the fractions.

1. $\frac{9}{45} = \frac{1}{5}$	2. $\frac{15}{20} = \frac{3}{4}$	3. $\frac{5}{15} = \frac{1}{3}$
4. $\frac{3}{18} = \frac{1}{6}$	5. $\frac{6}{18} = \frac{1}{3}$	6. $\frac{24}{30} = \frac{4}{5}$
7. $\frac{10}{15} = \frac{2}{3}$	8. $\frac{56}{80} = \frac{7}{10}$	9. $\frac{24}{56} = \frac{3}{7}$
10. $\frac{4}{36} = \frac{1}{9}$	11. $\frac{12}{20} = \frac{3}{5}$	12. $\frac{6}{15} = \frac{2}{5}$
13. $\frac{72}{126} = \frac{4}{7}$	14. $\frac{84}{144} = \frac{7}{12}$	15. $\frac{153}{171} = \frac{17}{19}$

PAGE 38

Simplifying Fractions

Simplify the fractions.

1. $\frac{5}{10} = \frac{1}{2}$	2. $\frac{16}{20} = \frac{4}{5}$	3. $\frac{18}{50} = \frac{9}{25}$
4. $\frac{24}{27} = \frac{8}{9}$	5. $\frac{8}{24} = \frac{1}{3}$	6. $\frac{28}{35} = \frac{4}{5}$
7. $\frac{12}{32} = \frac{3}{8}$	8. $\frac{9}{18} = \frac{1}{2}$	9. $\frac{20}{30} = \frac{2}{3}$
10. $\frac{15}{21} = \frac{5}{7}$	11. $\frac{18}{45} = \frac{2}{5}$	12. $\frac{14}{50} = \frac{7}{25}$
13. $\frac{64}{128} = \frac{1}{2}$	14. $\frac{75}{360} = \frac{5}{24}$	15. $\frac{125}{325} = \frac{5}{13}$

PAGE 39

Simplifying Fractions

Simplify the fractions.

1. $\frac{8}{28} = \frac{2}{7}$	2. $\frac{21}{35} = \frac{3}{5}$	3. $\frac{24}{32} = \frac{3}{4}$
4. $\frac{32}{40} = \frac{4}{5}$	5. $\frac{14}{35} = \frac{2}{5}$	6. $\frac{6}{20} = \frac{3}{10}$
7. $\frac{8}{40} = \frac{1}{5}$	8. $\frac{21}{42} = \frac{1}{2}$	9. $\frac{12}{75} = \frac{4}{25}$
10. $\frac{16}{36} = \frac{4}{9}$	11. $\frac{14}{22} = \frac{7}{11}$	12. $\frac{63}{72} = \frac{7}{8}$
13. $\frac{72}{168} = \frac{3}{7}$	14. $\frac{112}{154} = \frac{8}{11}$	15. $\frac{128}{256} = \frac{1}{2}$

PAGE 40

Improper Fractions to Mixed Numbers

Convert the improper fractions to mixed numbers.

1. $\frac{10}{9}$ — $1\frac{1}{9}$	2. $\frac{13}{3}$ — $4\frac{1}{3}$	3. $\frac{6}{5}$ — $1\frac{1}{5}$
4. $\frac{23}{4}$ — $5\frac{3}{4}$	5. $\frac{23}{9}$ — $2\frac{5}{9}$	6. $\frac{29}{7}$ — $4\frac{1}{7}$
7. $\frac{21}{5}$ — $4\frac{1}{5}$	8. $\frac{26}{7}$ — $3\frac{5}{7}$	9. $\frac{11}{4}$ — $2\frac{3}{4}$
10. $\frac{17}{4}$ — $4\frac{1}{4}$	11. $\frac{9}{4}$ — $2\frac{1}{4}$	12. $\frac{35}{6}$ — $5\frac{5}{6}$
13. $\frac{29}{12}$ — $2\frac{5}{12}$	14. $\frac{8}{7}$ — $1\frac{1}{7}$	15. $\frac{23}{10}$ — $2\frac{3}{10}$

PAGE 43

Improper Fractions to Mixed Numbers

Convert the improper fractions to mixed numbers.

1. $\frac{18}{7}$ $2\frac{4}{7}$	2. $\frac{12}{11}$ $1\frac{1}{11}$	3. $\frac{10}{7}$ $1\frac{3}{7}$
4. $\frac{38}{9}$ $4\frac{2}{9}$	5. $\frac{10}{9}$ $1\frac{1}{9}$	6. $\frac{43}{6}$ $7\frac{1}{6}$
7. $\frac{20}{11}$ $1\frac{9}{11}$	8. $\frac{16}{7}$ $2\frac{2}{7}$	9. $\frac{9}{8}$ $1\frac{1}{8}$
10. $\frac{5}{3}$ $1\frac{2}{3}$	11. $\frac{67}{12}$ $5\frac{7}{12}$	12. $\frac{11}{3}$ $3\frac{2}{3}$
13. $\frac{23}{8}$ $2\frac{7}{8}$	14. $\frac{63}{11}$ $5\frac{8}{11}$	15. $\frac{19}{7}$ $2\frac{5}{7}$

PAGE 44

Improper Fractions to Mixed Numbers

Convert the improper fractions to mixed numbers.

1. $\frac{5}{4}$ $1\frac{1}{4}$	2. $\frac{12}{7}$ $1\frac{5}{7}$	3. $\frac{16}{9}$ $1\frac{7}{9}$
4. $\frac{7}{4}$ $1\frac{3}{4}$	5. $\frac{13}{4}$ $3\frac{1}{4}$	6. $\frac{11}{10}$ $1\frac{1}{10}$
7. $\frac{22}{17}$ $1\frac{5}{17}$	8. $\frac{7}{5}$ $1\frac{2}{5}$	9. $\frac{25}{12}$ $2\frac{1}{12}$
10. $\frac{5}{2}$ $2\frac{1}{2}$	11. $\frac{59}{11}$ $5\frac{4}{11}$	12. $\frac{33}{10}$ $3\frac{3}{10}$
13. $\frac{14}{5}$ $2\frac{4}{5}$	14. $\frac{41}{12}$ $3\frac{5}{12}$	15. $\frac{15}{7}$ $2\frac{1}{7}$

PAGE 45

Improper Fractions to Mixed Numbers

Convert the improper fractions to mixed numbers.

1. $\frac{7}{6}$ $1\frac{1}{6}$	2. $\frac{41}{11}$ $3\frac{8}{11}$	3. $\frac{9}{5}$ $1\frac{4}{5}$
4. $\frac{8}{3}$ $2\frac{2}{3}$	5. $\frac{31}{8}$ $3\frac{7}{8}$	6. $\frac{24}{5}$ $4\frac{4}{5}$
7. $\frac{4}{3}$ $1\frac{1}{3}$	8. $\frac{13}{6}$ $2\frac{1}{6}$	9. $\frac{11}{2}$ $5\frac{1}{2}$
10. $\frac{31}{12}$ $2\frac{7}{12}$	11. $\frac{29}{11}$ $2\frac{7}{11}$	12. $\frac{46}{9}$ $5\frac{1}{9}$
13. $\frac{16}{3}$ $5\frac{1}{3}$	14. $\frac{28}{9}$ $3\frac{1}{9}$	15. $\frac{23}{5}$ $4\frac{3}{5}$

PAGE 46

Improper Fractions to Mixed Numbers

Convert the improper fractions to mixed numbers.

1. $\frac{33}{7}$ $4\frac{5}{7}$	2. $\frac{64}{11}$ $5\frac{9}{11}$	3. $\frac{37}{9}$ $4\frac{1}{9}$
4. $\frac{17}{3}$ $5\frac{2}{3}$	5. $\frac{19}{5}$ $3\frac{4}{5}$	6. $\frac{25}{6}$ $4\frac{1}{6}$
7. $\frac{29}{10}$ $2\frac{9}{10}$	8. $\frac{49}{12}$ $4\frac{1}{12}$	9. $\frac{9}{2}$ $4\frac{1}{2}$
10. $\frac{12}{5}$ $2\frac{2}{5}$	11. $\frac{39}{11}$ $3\frac{6}{11}$	12. $\frac{11}{7}$ $1\frac{4}{7}$
13. $\frac{25}{9}$ $2\frac{7}{9}$	14. $\frac{7}{2}$ $3\frac{1}{2}$	15. $\frac{19}{4}$ $4\frac{3}{4}$

PAGE 47

Improper Fractions to Mixed Numbers

Convert the improper fractions to mixed numbers.

1. $\frac{8}{5}$ — $1\frac{3}{5}$	2. $\frac{18}{5}$ — $3\frac{3}{5}$	3. $\frac{25}{11}$ — $2\frac{3}{11}$
4. $\frac{41}{8}$ — $5\frac{1}{8}$	5. $\frac{3}{2}$ — $1\frac{1}{2}$	6. $\frac{9}{7}$ — $1\frac{2}{7}$
7. $\frac{31}{11}$ — $2\frac{9}{11}$	8. $\frac{27}{7}$ — $3\frac{6}{7}$	9. $\frac{10}{3}$ — $3\frac{1}{3}$
10. $\frac{26}{9}$ — $2\frac{8}{9}$	11. $\frac{52}{11}$ — $4\frac{8}{11}$	12. $\frac{14}{9}$ — $1\frac{5}{9}$
13. $\frac{49}{10}$ — $4\frac{9}{10}$	14. $\frac{17}{9}$ — $1\frac{8}{9}$	15. $\frac{23}{6}$ — $3\frac{5}{6}$

PAGE 48

Improper Fractions to Mixed Numbers

Convert the mixed numbers to improper fractions.

1. $4\frac{1}{6}$ — $\frac{25}{6}$	2. $3\frac{2}{7}$ — $\frac{23}{7}$	3. $2\frac{7}{8}$ — $\frac{23}{8}$
4. $6\frac{1}{4}$ — $\frac{25}{4}$	5. $1\frac{7}{9}$ — $\frac{16}{9}$	6. $5\frac{5}{8}$ — $\frac{45}{8}$
7. $2\frac{2}{5}$ — $\frac{12}{5}$	8. $3\frac{9}{11}$ — $\frac{42}{11}$	9. $8\frac{3}{5}$ — $\frac{43}{5}$
10. $3\frac{7}{8}$ — $\frac{31}{8}$	11. $1\frac{2}{3}$ — $\frac{5}{3}$	12. $5\frac{7}{9}$ — $\frac{52}{9}$
13. $1\frac{4}{5}$ — $\frac{9}{5}$	14. $2\frac{7}{10}$ — $\frac{27}{10}$	15. $7\frac{4}{5}$ — $\frac{39}{5}$

PAGE 50

Improper Fractions to Mixed Numbers

Convert the mixed numbers to improper fractions.

1. $5\frac{6}{11}$ — $\frac{61}{11}$	2. $4\frac{7}{9}$ — $\frac{43}{9}$	3. $2\frac{9}{10}$ — $\frac{29}{10}$
4. $2\frac{1}{6}$ — $\frac{13}{6}$	5. $1\frac{9}{11}$ — $\frac{20}{11}$	6. $2\frac{1}{7}$ — $\frac{15}{7}$
7. $4\frac{1}{2}$ — $\frac{9}{2}$	8. $3\frac{3}{7}$ — $\frac{24}{7}$	9. $5\frac{4}{5}$ — $\frac{29}{5}$
10. $1\frac{2}{7}$ — $\frac{9}{7}$	11. $1\frac{1}{4}$ — $\frac{5}{4}$	12. $4\frac{3}{5}$ — $\frac{23}{5}$
13. $5\frac{3}{4}$ — $\frac{23}{4}$	14. $8\frac{2}{3}$ — $\frac{26}{3}$	15. $2\frac{1}{4}$ — $\frac{9}{4}$

PAGE 51

Improper Fractions to Mixed Numbers

Convert the mixed numbers to improper fractions.

1. $5\frac{1}{8}$ — $\frac{41}{8}$	2. $1\frac{8}{9}$ — $\frac{17}{9}$	3. $3\frac{1}{3}$ — $\frac{10}{3}$
4. $3\frac{7}{9}$ — $\frac{34}{9}$	5. $5\frac{7}{10}$ — $\frac{57}{10}$	6. $2\frac{1}{5}$ — $\frac{11}{5}$
7. $4\frac{3}{4}$ — $\frac{19}{4}$	8. $7\frac{1}{5}$ — $\frac{36}{5}$	9. $4\frac{4}{7}$ — $\frac{32}{7}$
10. $4\frac{2}{3}$ — $\frac{14}{3}$	11. $3\frac{4}{5}$ — $\frac{19}{5}$	12. $1\frac{3}{8}$ — $\frac{11}{8}$
13. $2\frac{5}{8}$ — $\frac{21}{8}$	14. $4\frac{5}{11}$ — $\frac{49}{11}$	15. $8\frac{9}{10}$ — $\frac{89}{10}$

PAGE 52

Improper Fractions to Mixed Numbers

Convert the mixed numbers to improper fractions.

1. $1\frac{1}{7}$ = $\frac{8}{7}$	2. $4\frac{2}{5}$ = $\frac{22}{5}$	3. $1\frac{1}{2}$ = $\frac{3}{2}$
4. $3\frac{1}{5}$ = $\frac{16}{5}$	5. $2\frac{5}{6}$ = $\frac{17}{6}$	6. $4\frac{7}{11}$ = $\frac{51}{11}$
7. $5\frac{9}{10}$ = $\frac{59}{10}$	8. $7\frac{2}{3}$ = $\frac{23}{3}$	9. $3\frac{5}{12}$ = $\frac{41}{12}$
10. $6\frac{4}{5}$ = $\frac{34}{5}$	11. $1\frac{1}{10}$ = $\frac{11}{10}$	12. $5\frac{1}{3}$ = $\frac{16}{3}$
13. $5\frac{3}{5}$ = $\frac{28}{5}$	14. $2\frac{8}{9}$ = $\frac{26}{9}$	15. $4\frac{5}{6}$ = $\frac{29}{6}$

PAGE 53

Improper Fractions to Mixed Numbers

Convert the mixed numbers to improper fractions.

1. $4\frac{7}{8}$ = $\frac{39}{8}$	2. $4\frac{1}{7}$ = $\frac{29}{7}$	3. $5\frac{5}{7}$ = $\frac{40}{7}$
4. $2\frac{2}{5}$ = $\frac{12}{5}$	5. $6\frac{5}{8}$ = $\frac{53}{8}$	6. $3\frac{6}{7}$ = $\frac{27}{7}$
7. $7\frac{1}{2}$ = $\frac{15}{2}$	8. $3\frac{2}{3}$ = $\frac{11}{3}$	9. $5\frac{2}{9}$ = $\frac{47}{9}$
10. $4\frac{8}{9}$ = $\frac{44}{9}$	11. $8\frac{1}{2}$ = $\frac{17}{2}$	12. $5\frac{1}{4}$ = $\frac{21}{4}$
13. $1\frac{3}{4}$ = $\frac{7}{4}$	14. $2\frac{2}{3}$ = $\frac{8}{3}$	15. $3\frac{3}{8}$ = $\frac{27}{8}$

PAGE 54

Fraction Addition and Subtraction

Add the fractions and simplify. *(Remember to convert improper fractions to mixed numbers.)*

1. $\frac{2}{3} + \frac{1}{5} = \frac{13}{15}$
2. $\frac{1}{4} + \frac{1}{7} = \frac{11}{28}$
3. $\frac{5}{9} + \frac{1}{2} = 1\frac{1}{18}$
4. $\frac{1}{4} + \frac{2}{3} = \frac{11}{12}$
5. $\frac{1}{2} + \frac{5}{7} = 1\frac{3}{14}$

Subtract the fractions and simplify.

6. $\frac{7}{8} - \frac{1}{4} = \frac{5}{8}$
7. $\frac{9}{10} - \frac{2}{3} = \frac{7}{30}$
8. $\frac{1}{2} - \frac{2}{5} = \frac{1}{10}$
9. $\frac{2}{3} - \frac{4}{7} = \frac{2}{21}$
10. $\frac{6}{7} - \frac{1}{4} = \frac{17}{28}$

PAGE 57

Fraction Addition and Subtraction

Add the fractions and simplify. *(Remember to convert improper fractions to mixed numbers.)*

1. $\frac{1}{4} + \frac{1}{3} = \frac{7}{12}$
2. $\frac{9}{11} + \frac{1}{2} = 1\frac{7}{22}$
3. $\frac{2}{3} + \frac{5}{9} = 1\frac{2}{9}$
4. $\frac{9}{10} + \frac{1}{4} = 1\frac{3}{20}$
5. $\frac{1}{3} + \frac{5}{7} = 1\frac{1}{21}$

Subtract the fractions and simplify.

6. $\frac{3}{4} - \frac{1}{2} = \frac{1}{4}$
7. $\frac{3}{4} - \frac{1}{7} = \frac{17}{28}$
8. $\frac{1}{2} - \frac{4}{9} = \frac{1}{18}$
9. $\frac{7}{8} - \frac{2}{3} = \frac{5}{24}$
10. $\frac{2}{3} - \frac{3}{5} = \frac{1}{15}$

PAGE 58

Fraction Addition and Subtraction

Add the fractions and simplify. *(Remember to convert improper fractions to mixed numbers.)*

1. $\frac{9}{10} + \frac{3}{4} = 1\frac{13}{20}$
2. $\frac{4}{7} + \frac{4}{5} = 1\frac{13}{35}$
3. $\frac{1}{2} + \frac{7}{8} = 1\frac{3}{8}$
4. $\frac{3}{4} + \frac{2}{3} = 1\frac{5}{12}$
5. $\frac{2}{5} + \frac{3}{10} = \frac{7}{10}$

Subtract the fractions and simplify.

6. $\frac{3}{5} - \frac{1}{3} = \frac{4}{15}$
7. $\frac{5}{8} - \frac{2}{5} = \frac{9}{40}$
8. $\frac{8}{9} - \frac{2}{3} = \frac{2}{9}$
9. $\frac{3}{5} - \frac{1}{10} = \frac{1}{2}$
10. $\frac{1}{4} - \frac{1}{5} = \frac{1}{20}$

PAGE 59

Fraction Addition and Subtraction

Add the fractions and simplify. *(Remember to convert improper fractions to mixed numbers.)*

1. $\frac{2}{3} + \frac{1}{6} = \frac{5}{6}$
2. $\frac{3}{4} + \frac{1}{9} = \frac{31}{36}$
3. $\frac{1}{3} + \frac{3}{4} = 1\frac{1}{12}$
4. $\frac{1}{2} + \frac{9}{10} = 1\frac{2}{5}$
5. $\frac{4}{11} + \frac{2}{3} = 1\frac{1}{33}$

Subtract the fractions and simplify.

6. $\frac{4}{9} - \frac{2}{6} = \frac{1}{9}$
7. $\frac{5}{9} - \frac{1}{12} = \frac{17}{36}$
8. $\frac{5}{8} - \frac{1}{4} = \frac{3}{8}$
9. $\frac{3}{4} - \frac{2}{5} = \frac{7}{20}$
10. $\frac{8}{11} - \frac{4}{7} = \frac{12}{77}$

PAGE 60

Fraction Addition and Subtraction

Add the fractions and simplify. *(Remember to convert improper fractions to mixed numbers.)*

1. $\frac{2}{5} + \frac{1}{5} = \frac{3}{5}$
2. $\frac{4}{9} + \frac{2}{3} = 1\frac{1}{9}$
3. $\frac{1}{4} + \frac{2}{5} = \frac{13}{20}$
4. $\frac{1}{2} + \frac{4}{9} = \frac{17}{18}$
5. $\frac{6}{7} + \frac{1}{4} = 1\frac{3}{28}$

Subtract the fractions and simplify.

6. $\frac{2}{3} - \frac{1}{6} = \frac{1}{2}$
7. $\frac{3}{4} - \frac{4}{9} = \frac{11}{36}$
8. $\frac{5}{6} - \frac{4}{5} = \frac{1}{30}$
9. $\frac{7}{9} - \frac{1}{3} = \frac{4}{9}$
10. $\frac{6}{7} - \frac{3}{5} = \frac{9}{35}$

PAGE 61

Fraction Addition and Subtraction

Add the fractions and simplify. *(Remember to convert improper fractions to mixed numbers.)*

1. $\frac{3}{5} + \frac{1}{6} = \frac{23}{30}$
2. $\frac{1}{3} + \frac{8}{9} = 1\frac{2}{9}$
3. $\frac{5}{6} + \frac{2}{5} = 1\frac{7}{30}$
4. $\frac{2}{9} + \frac{1}{12} = \frac{11}{36}$
5. $\frac{1}{3} + \frac{5}{6} = 1\frac{1}{6}$

Subtract the fractions and simplify.

6. $\frac{3}{4} - \frac{1}{7} = \frac{17}{28}$
7. $\frac{1}{3} - \frac{1}{9} = \frac{2}{9}$
8. $\frac{3}{5} - \frac{3}{7} = \frac{6}{35}$
9. $\frac{3}{4} - \frac{2}{3} = \frac{1}{12}$
10. $\frac{1}{5} - \frac{1}{6} = \frac{1}{30}$

PAGE 62

Fraction Addition and Subtraction

Add the fractions and simplify. *(Remember to convert improper fractions to mixed numbers.)*

1. $\frac{3}{5} + \frac{1}{2} = 1\frac{1}{10}$
2. $\frac{2}{3} + \frac{1}{9} = \frac{7}{9}$
3. $\frac{4}{5} + \frac{3}{7} = 1\frac{8}{35}$
4. $\frac{7}{9} + \frac{2}{3} = 1\frac{4}{9}$
5. $\frac{3}{8} + \frac{1}{6} = \frac{13}{24}$

Subtract the fractions and simplify.

6. $\frac{5}{6} - \frac{2}{3} = \frac{1}{6}$
7. $\frac{5}{7} - \frac{1}{2} = \frac{3}{14}$
8. $\frac{1}{2} - \frac{1}{6} = \frac{1}{3}$
9. $\frac{2}{3} - \frac{5}{8} = \frac{1}{24}$
10. $\frac{9}{10} - \frac{3}{4} = \frac{3}{20}$

PAGE 63

Fraction Addition and Subtraction

Add the fractions and simplify. *(Remember to convert improper fractions to mixed numbers.)*

1. $\frac{5}{9} + \frac{1}{6} = \frac{13}{18}$
2. $\frac{1}{7} + \frac{1}{2} = \frac{9}{14}$
3. $\frac{2}{3} + \frac{6}{7} = 1\frac{11}{21}$
4. $\frac{1}{2} + \frac{8}{9} = 1\frac{7}{18}$
5. $\frac{1}{4} + \frac{1}{9} = \frac{13}{36}$

Subtract the fractions and simplify.

6. $\frac{4}{11} - \frac{1}{3} = \frac{1}{33}$
7. $\frac{6}{7} - \frac{4}{5} = \frac{2}{35}$
8. $\frac{3}{4} - \frac{1}{8} = \frac{5}{8}$
9. $\frac{1}{2} - \frac{3}{7} = \frac{1}{14}$
10. $\frac{4}{5} - \frac{3}{4} = \frac{1}{20}$

PAGE 64

Addition and Subtraction with Mixed Numbers

Add the fractions and simplify.

1. $1\frac{1}{3} + 2\frac{1}{3} = 3\frac{2}{3}$
2. $5\frac{1}{4} + 3\frac{2}{4} = 8\frac{3}{4}$
3. $1\frac{1}{2} + 1\frac{1}{4} = 2\frac{3}{4}$
4. $1\frac{1}{5} + 2\frac{1}{2} = 3\frac{7}{10}$
5. $1\frac{1}{2} + 1\frac{2}{3} = 3\frac{1}{6}$

Subtract the fractions and simplify.

6. $2\frac{1}{5} - 1\frac{2}{5} = \frac{4}{5}$
7. $3\frac{1}{3} - 2\frac{2}{3} = \frac{2}{3}$
8. $1\frac{1}{2} - 1\frac{1}{3} = \frac{1}{6}$
9. $2\frac{1}{3} - 1\frac{1}{4} = 1\frac{1}{12}$
10. $3\frac{1}{2} - 2\frac{1}{5} = 1\frac{3}{10}$

PAGE 68

Addition and Subtraction with Mixed Numbers

Add the fractions and simplify.

1. $1\frac{3}{5} + 4\frac{1}{5} = 5\frac{4}{5}$
2. $2\frac{4}{7} + 3\frac{5}{7} = 6\frac{2}{7}$
3. $2\frac{1}{2} + 2\frac{1}{4} = 4\frac{3}{4}$
4. $1\frac{1}{8} + 1\frac{1}{2} = 2\frac{5}{8}$
5. $2\frac{1}{5} + 1\frac{1}{10} = 3\frac{3}{10}$

Subtract the fractions and simplify.

6. $3\frac{1}{4} - 1\frac{3}{4} = 1\frac{1}{2}$
7. $2\frac{1}{3} - 1\frac{2}{3} = \frac{2}{3}$
8. $2\frac{1}{4} - 1\frac{1}{8} = 1\frac{1}{8}$
9. $1\frac{2}{3} - 1\frac{1}{2} = \frac{1}{6}$
10. $2\frac{1}{5} - 1\frac{1}{2} = \frac{7}{10}$

PAGE 69

Addition and Subtraction with Mixed Numbers

Add the fractions and simplify.

1. $1\frac{1}{8} + 3\frac{5}{8} = 4\frac{3}{4}$
2. $2\frac{1}{5} + 2\frac{2}{5} = 4\frac{3}{5}$
3. $3\frac{1}{2} + 1\frac{2}{3} = 5\frac{1}{6}$
4. $1\frac{1}{3} + 1\frac{3}{4} = 3\frac{1}{12}$
5. $1\frac{1}{4} + 1\frac{1}{5} = 2\frac{9}{20}$

Subtract the fractions and simplify.

6. $3\frac{2}{5} - 1\frac{3}{5} = 1\frac{4}{5}$
7. $4\frac{1}{2} - 2\frac{1}{2} = 2$
8. $2\frac{1}{8} - 1\frac{1}{4} = \frac{7}{8}$
9. $1\frac{1}{2} - 1\frac{2}{5} = \frac{1}{10}$
10. $3\frac{1}{4} - 1\frac{1}{3} = 1\frac{11}{12}$

PAGE 70

Addition and Subtraction with Mixed Numbers

Add the fractions and simplify.

1. $2\frac{1}{6} + 1\frac{1}{6} = 3\frac{1}{3}$
2. $1\frac{2}{3} + 2\frac{1}{3} = 4$
3. $1\frac{1}{2} + 3\frac{1}{3} = 4\frac{5}{6}$
4. $1\frac{1}{3} + 1\frac{2}{9} = 2\frac{5}{9}$
5. $2\frac{1}{4} + 1\frac{1}{8} = 3\frac{3}{8}$

Subtract the fractions and simplify.

6. $2\frac{3}{8} - 1\frac{1}{8} = 1\frac{1}{4}$
7. $3\frac{1}{7} - 2\frac{3}{7} = \frac{5}{7}$
8. $3\frac{1}{2} - 1\frac{2}{8} = 2\frac{1}{4}$
9. $2\frac{1}{5} - 1\frac{1}{4} = \frac{19}{20}$
10. $3\frac{3}{4} - 2\frac{1}{2} = 1\frac{1}{4}$

PAGE 71

Addition and Subtraction with Mixed Numbers

Add the fractions and simplify.

1. $1\frac{2}{7} + 3\frac{4}{7} = 4\frac{6}{7}$
2. $1\frac{3}{5} + 1\frac{4}{5} = 3\frac{2}{5}$
3. $1\frac{2}{5} + 1\frac{1}{2} = 2\frac{9}{10}$
4. $2\frac{1}{2} + 1\frac{1}{4} = 3\frac{3}{4}$
5. $1\frac{1}{2} + 2\frac{1}{3} = 3\frac{5}{6}$

Subtract the fractions and simplify.

6. $2\frac{1}{5} - 1\frac{3}{5} = \frac{3}{5}$
7. $3\frac{1}{8} - 2\frac{3}{8} = \frac{3}{4}$
8. $1\frac{3}{4} - 1\frac{1}{5} = \frac{11}{20}$
9. $2\frac{1}{3} - 1\frac{1}{2} = \frac{5}{6}$
10. $3\frac{1}{2} - 2\frac{1}{8} = 1\frac{3}{8}$

PAGE 72

Addition and Subtraction with Mixed Numbers

Add the fractions and simplify.

1. $2\frac{1}{8} + 3\frac{3}{8} = 5\frac{1}{2}$
2. $1\frac{5}{9} + 3\frac{7}{9} = 5\frac{1}{3}$
3. $1\frac{1}{3} + 1\frac{1}{4} = 2\frac{7}{12}$
4. $1\frac{1}{2} + 1\frac{1}{3} = 2\frac{5}{6}$
5. $1\frac{2}{3} + 2\frac{1}{4} = 3\frac{11}{12}$

Subtract the fractions and simplify.

6. $2\frac{1}{6} - 1\frac{5}{6} = \frac{1}{3}$
7. $4\frac{2}{3} - 2\frac{1}{3} = 2\frac{1}{3}$
8. $5\frac{1}{2} - 3\frac{1}{4} = 2\frac{1}{4}$
9. $2\frac{3}{5} - 1\frac{1}{2} = 1\frac{1}{10}$
10. $3\frac{1}{2} - 2\frac{1}{10} = 1\frac{2}{5}$

PAGE 73

Addition and Subtraction with Mixed Numbers

Add the fractions and simplify.

1. $4\frac{1}{4} + 2\frac{2}{4} = 6\frac{3}{4}$
2. $2\frac{1}{10} + 3\frac{3}{10} = 5\frac{2}{5}$
3. $1\frac{1}{3} + 2\frac{1}{4} = 3\frac{7}{12}$
4. $2\frac{1}{3} + 1\frac{1}{6} = 3\frac{1}{2}$
5. $1\frac{2}{5} + 1\frac{1}{10} = 2\frac{1}{2}$

Subtract the fractions and simplify.

6. $3\frac{1}{5} - 2\frac{2}{5} = \frac{4}{5}$
7. $2\frac{3}{4} - 1\frac{1}{4} = 1\frac{1}{2}$
8. $4\frac{1}{5} - 1\frac{1}{10} = 3\frac{1}{10}$
9. $3\frac{1}{4} - 2\frac{1}{2} = \frac{3}{4}$
10. $2\frac{5}{8} - 1\frac{1}{4} = 1\frac{3}{8}$

PAGE 74

Fraction Muliplication and Division

Multiply the fractions and simplify.

1. $\frac{6}{7} \times \frac{2}{3} = \frac{4}{7}$
2. $\frac{2}{7} \times \frac{7}{12} = \frac{1}{6}$
3. $\frac{3}{4} \times \frac{7}{9} = \frac{7}{12}$
4. $\frac{2}{3} \times \frac{7}{10} = \frac{7}{15}$
5. $\frac{1}{2} \times \frac{7}{11} = \frac{7}{22}$

Divide the fractions and simplify.

6. $\frac{1}{6} \div \frac{1}{2} = \frac{1}{3}$
7. $\frac{4}{5} \div \frac{3}{8} = 2\frac{2}{15}$
8. $\frac{1}{2} \div \frac{8}{11} = \frac{11}{16}$
9. $\frac{2}{5} \div \frac{1}{5} = 2$
10. $\frac{3}{8} \div 2 = \frac{3}{16}$

PAGE 77

Fraction Muliplication and Division

Multiply the fractions and simplify.

1. $\frac{8}{11} \times \frac{3}{4} = \frac{6}{11}$
2. $\frac{2}{3} \times \frac{5}{6} = \frac{5}{9}$
3. $\frac{1}{2} \times \frac{3}{4} = \frac{3}{8}$
4. $\frac{1}{2} \times \frac{2}{3} = \frac{1}{3}$
5. $\frac{4}{5} \times \frac{7}{10} = \frac{14}{25}$

Divide the fractions and simplify.

6. $\frac{3}{4} \div \frac{5}{9} = 1\frac{7}{20}$
7. $\frac{3}{10} \div \frac{9}{10} = \frac{1}{3}$
8. $\frac{4}{11} \div 4 = \frac{1}{11}$
9. $\frac{3}{5} \div \frac{4}{5} = \frac{3}{4}$
10. $\frac{2}{3} \div \frac{3}{8} = 1\frac{7}{9}$

PAGE 78

Fraction Muliplication and Division

Multiply the fractions and simplify.

1. $\frac{5}{12} \times \frac{1}{2} = \frac{5}{24}$
2. $\frac{3}{7} \times \frac{1}{6} = \frac{1}{14}$
3. $\frac{1}{3} \times \frac{1}{2} = \frac{1}{6}$
4. $\frac{6}{11} \times \frac{1}{4} = \frac{3}{22}$
5. $\frac{4}{5} \times \frac{4}{5} = \frac{16}{25}$

Divide the fractions and simplify.

6. $\frac{3}{10} \div \frac{2}{3} = \frac{9}{20}$
7. $\frac{5}{7} \div \frac{5}{6} = \frac{6}{7}$
8. $\frac{5}{8} \div 5 = \frac{1}{8}$
9. $\frac{7}{9} \div \frac{5}{12} = 1\frac{13}{15}$
10. $\frac{1}{2} \div \frac{3}{4} = \frac{2}{3}$

PAGE 79

Fraction Muliplication and Division

Multiply the fractions and simplify.

1. $\frac{9}{10} \times \frac{1}{3} = \frac{3}{10}$
2. $\frac{5}{6} \times \frac{9}{10} = \frac{3}{4}$
3. $\frac{3}{7} \times \frac{7}{12} = \frac{1}{4}$
4. $\frac{5}{6} \times \frac{3}{4} = \frac{5}{8}$
5. $\frac{2}{3} \times \frac{1}{5} = \frac{2}{15}$

Divide the fractions and simplify.

6. $\frac{7}{11} \div \frac{5}{11} = 1\frac{2}{5}$
7. $\frac{5}{9} \div 4 = \frac{5}{36}$
8. $\frac{2}{3} \div \frac{3}{4} = \frac{8}{9}$
9. $\frac{3}{4} \div \frac{1}{2} = 1\frac{1}{2}$
10. $\frac{1}{6} \div \frac{2}{3} = \frac{1}{4}$

PAGE 80

Fraction Muliplication and Division

Multiply the fractions and simplify.

1. $\frac{5}{12} \times \frac{6}{7} = \frac{5}{14}$
2. $\frac{3}{4} \times \frac{1}{5} = \frac{3}{20}$
3. $\frac{4}{7} \times \frac{1}{2} = \frac{2}{7}$
4. $\frac{3}{5} \times \frac{2}{3} = \frac{2}{5}$
5. $\frac{2}{7} \times \frac{3}{4} = \frac{3}{14}$

Divide the fractions and simplify.

6. $\frac{3}{5} \div \frac{9}{10} = \frac{2}{3}$
7. $\frac{4}{5} \div \frac{3}{7} = 1\frac{13}{15}$
8. $\frac{6}{11} \div \frac{1}{2} = 1\frac{1}{11}$
9. $\frac{5}{9} \div 3 = \frac{5}{27}$
10. $\frac{2}{3} \div \frac{4}{7} = 1\frac{1}{6}$

PAGE 81

Fraction Muliplication and Division

Multiply the fractions and simplify.

1. $\frac{3}{4} \times \frac{6}{11} = \frac{9}{22}$
2. $\frac{1}{5} \times \frac{4}{5} = \frac{4}{25}$
3. $\frac{5}{8} \times \frac{2}{3} = \frac{5}{12}$
4. $\frac{2}{9} \times \frac{3}{4} = \frac{1}{6}$
5. $\frac{8}{11} \times \frac{1}{2} = \frac{4}{11}$

Divide the fractions and simplify.

6. $\frac{4}{5} \div 2 = \frac{2}{5}$
7. $\frac{3}{4} \div \frac{2}{3} = 1\frac{1}{8}$
8. $\frac{1}{3} \div \frac{1}{5} = 1\frac{2}{3}$
9. $\frac{5}{12} \div \frac{1}{2} = \frac{5}{6}$
10. $\frac{2}{3} \div \frac{5}{12} = 1\frac{3}{5}$

PAGE 82

Fraction Muliplication and Division

Multiply the fractions and simplify.

1. $\frac{4}{9} \times \frac{9}{10} = \frac{2}{5}$
2. $\frac{1}{4} \times \frac{1}{4} = \frac{1}{16}$
3. $\frac{1}{6} \times \frac{3}{4} = \frac{1}{8}$
4. $\frac{7}{12} \times \frac{3}{4} = \frac{7}{16}$
5. $\frac{1}{2} \times \frac{3}{5} = \frac{3}{10}$

Divide the fractions and simplify.

6. $\frac{9}{10} \div \frac{1}{3} = 2\frac{7}{10}$
7. $\frac{1}{4} \div \frac{2}{3} = \frac{3}{8}$
8. $\frac{6}{7} \div 3 = \frac{2}{7}$
9. $\frac{1}{2} \div \frac{9}{10} = \frac{5}{9}$
10. $\frac{2}{7} \div \frac{2}{5} = \frac{5}{7}$

PAGE 83

Fraction Muliplication and Division

Multiply the fractions and simplify.

1. $\frac{7}{10} \times \frac{6}{7} = \frac{3}{5}$
2. $\frac{7}{8} \times \frac{4}{5} = \frac{7}{10}$
3. $\frac{4}{5} \times \frac{2}{3} = \frac{8}{15}$
4. $\frac{2}{9} \times \frac{1}{4} = \frac{1}{18}$
5. $\frac{2}{3} \times \frac{9}{11} = \frac{6}{11}$

Divide the fractions and simplify.

6. $\frac{4}{7} \div \frac{1}{2} = 1\frac{1}{7}$
7. $\frac{5}{6} \div \frac{1}{3} = 2\frac{1}{2}$
8. $\frac{1}{2} \div 2 = \frac{1}{4}$
9. $\frac{7}{9} \div \frac{7}{8} = \frac{8}{9}$
10. $\frac{3}{8} \div \frac{3}{10} = 1\frac{1}{4}$

PAGE 84

Fraction Muliplication and Division

Multiply the fractions and simplify.

1. $\frac{7}{8} \times \frac{8}{9} = \frac{7}{9}$
2. $\frac{2}{3} \times \frac{6}{11} = \frac{4}{11}$
3. $\frac{1}{4} \times \frac{1}{3} = \frac{1}{12}$
4. $\frac{1}{2} \times \frac{5}{9} = \frac{5}{18}$
5. $\frac{5}{9} \times \frac{3}{5} = \frac{1}{3}$

Divide the fractions and simplify.

6. $\frac{3}{5} \div \frac{5}{6} = \frac{18}{25}$
7. $\frac{1}{2} \div \frac{6}{7} = \frac{7}{12}$
8. $\frac{7}{10} \div \frac{1}{2} = 1\frac{2}{5}$
9. $\frac{2}{3} \div 4 = \frac{1}{6}$
10. $\frac{5}{6} \div \frac{3}{4} = 1\frac{1}{9}$

PAGE 85

Fraction Muliplication and Division

Multiply the fractions and simplify.

1. $\frac{1}{2} \times \frac{5}{6} = \frac{5}{12}$
2. $\frac{5}{7} \times \frac{1}{3} = \frac{5}{21}$
3. $\frac{3}{8} \times \frac{8}{9} = \frac{1}{3}$
4. $\frac{3}{4} \times \frac{4}{5} = \frac{3}{5}$
5. $\frac{1}{6} \times \frac{9}{11} = \frac{3}{22}$

Divide the fractions and simplify.

6. $\frac{1}{3} \div 3 = \frac{1}{9}$
7. $\frac{1}{2} \div \frac{3}{6} = 1$
8. $\frac{5}{6} \div \frac{1}{2} = 1\frac{2}{3}$
9. $\frac{2}{3} \div \frac{6}{7} = \frac{7}{9}$
10. $\frac{7}{10} \div \frac{2}{3} = 1\frac{1}{20}$

PAGE 86

Next Step:

Boost your math skills even further using our Pre-Algebra Skill Building Workbook. This instructional workbook features step-by-step explanations and examples with lots of practice worksheets - answers included.

Pre-Algebra Skill Building Workbook:
with Explanations, Examples, Practice Problems and Answers

Some Topics include:

Negative Numbers
Exponents & Roots
Factors & Multiples
Fractions
Decimals & Percents
Expressions & Operations
Equations
Graphing

More Fraction Practice:

For extra practice with fractions to perfect your skills, try our Big Book of Math Practice Problems: Fractions and Decimals.

Big Book of Math Practice Problems:
Fractions & Decimals

Made in the USA
Las Vegas, NV
20 March 2026

43997548R00063